Cesar Lattes, a descoberta do méson π e outras histórias

2ª Edição

Edição comemorativa do centenário de nascimento de Lattes

Cesar Lattes, a descoberta do méson π e outras histórias

2ª Edição

Edição comemorativa do centenário de nascimento de Lattes

Organizadores

F. Caruso, A. Marques (*in memoriam*) & A. Troper

Copyright © 2024 os organizadores
2ª Edição

Direção editorial: Victor Pereira Marinho e José Roberto Marinho

Capa: Fabrício Ribeiro
Projeto gráfico e diagramação: Francisco Caruso

Edição revisada segundo o Novo Acordo Ortográfico da Língua Portuguesa

Dados Internacionais de Catalogação na publicação (CIP)
(Câmara Brasileira do Livro, SP, Brasil)

Cesar Lattes, a descoberta do méson π e outras histórias / organizadores F. Caruso, A. Marques (in memoriam), A. Troper. – 2. ed. – São Paulo: LF Editorial, 2024.

Edição trilíngue: português/inglês/espanhol.
Bibliografia
ISBN 978-65-5563-440-2

1. Físicos - Brasil - Autobiografia 2. Lattes, Cesare Mansueto Giulio, 1924-2005 I. Caruso, F. II. Marques, A. III. Troper, A.

24-200054 CDD-530.092

Índices para catálogo sistemático:
1. Físicos: Vida e obra 530.092

Tábata Alves da Silva - Bibliotecária - CRB-8/9253

Todos os direitos reservados. Nenhuma parte desta obra poderá ser reproduzida sejam quais forem os meios empregados sem a permissão da Editora. Aos infratores aplicam-se as sanções previstas nos artigos 102, 104, 106 e 107 da Lei Nº 9.610, de 19 de fevereiro de 1998

LF Editorial
www.livrariadafisica.com.br
www.lfeditorial.com.br
(11) 2648-6666 | Loja do Instituto de Física da USP
(11) 3936-3413 | Editora

Sumário

Francisco Caruso & Amós Troper

1 Prefácio à segunda edição   1

Israel Vargas

2 Prefácio   3

Francisco Caruso, Alfredo Marques & Amós Troper

3 Nota dos Editores   5

I. TESTEMUNHOS DE PRECURSORES, PARTICIPANTES E CONTEMPORÂNEOS À DESCOBERTA DO MÉSON π 9

Gleb Wataghin

4 Saudações a Cesar Lattes por ocasião de seus 60 anos   9

Giuseppe Occhialini

5 Cesar Lattes: Bristol Years   11

Cesar Lattes

6 My Work in Meson Physics with Nuclear Emulsions   15

George Dixon Rochester

7 The Development and use of Nuclear Emulsions in England in the Years 1945-50   19

Owen Lock

8 Cesar Lattes Half a Century Ago: The Pion Pioneers   23

Donald Perkins

9 The Discovery of the Pion in Bristol in 1947   27

Yoichi Fujimoto
10 Discovery of Pi and Mu Mesons and Brazil-Japan Collaboration on Cosmic-Rays
39

II. A DESCOBERTA DO MÉSON π NOS RELATOS DA POS-TERIDADE
51

José Maria Filardo Bassalo
11 César Lattes: um dos descobridores do méson π
51

Edison Hiroyuki Shibuya
12 César Lattes
67

Ana Maria Ribeiro de Andrade
13 O Sucesso do Grupo de Bristol
73

Antonio Augusto Passos Videira & Cássio Leite Vieira
14 Os Físicos e suas Histórias: os Usos e as Versões da Detecção do Méson PI
93

III. OUTRAS HISTÓRIAS
107

Cesar Lattes
15 O Nascimento das Partículas Elementares
107

Carlos Aguirre Bastos
16 Formación del Laboratorio de Física Cósmica de Chacaltaya
133

Alfredo Marques
17 CBPF: da Descoberta do Méson π aos Dez Primeiros Anos
159

Alberto Santoro
18 Cosmic Ray Physics and the Very Large Hadron Collider
231

Cássio Leite Vieira
19 O Méson pi em Bristol Hoje
235

1

Prefácio à segunda edição

Francisco Caruso & Amós Troper

A figura de Cesar Lattes estará para sempre ligada a um novo capítulo na história da Física de Partículas, ao desenvolvimento e institucionalização da pesquisa científica no Brasil, incluindo as criações do Centro Brasileiro de Pesquisas Físicas e o CNPq. Sendo assim, tem, ao longo dos anos, recebido justas homenagens acadêmicas país afora. Em particular, no CBPF, como parte dos festejos de seus 70 anos de vida, Alfredo Marques editou e publicou o livro *Cesar Lattes – 70 Anos: A Nova Física Brasileira*, Rio de Janeiro: CBPF, em 1994. Em seguida, por ocasião do cinquentenário desta instituição, houve vários tributos a Lattes. Alfredo publicou um livro paradidático chamado *A descoberta do méson* π, destinado e distribuído aos jovens que então participavam do Programa de Vocação Científica, PROVOC, do CBPF. Com a colaboração de Caruso e Troper, foi também impresso o livro *Cesar Lattes, a descoberta do méson* π *e outras histórias*, Rio de Janeiro: CBPF, 1999, com uma pequena tiragem de 500 exemplares, que logo se esgotou. Por fim, foi produzido um selo autoadesivo comemorativo à descoberta do méson π, que foi amplamente utilizado nas publicações e documentos do CBPF naquele ano. Ademais, em 2018, saiu um número especial da revista *Ciência e Sociedade* (vol. 5, n. 3), em comemoração aos 70 Anos do Méson π.

No próximo dia 11 de julho, Lattes completaria 100 anos de vida. A publicação dessa segunda edição revista e corrigida de *Cesar Lattes, a descoberta do méson* π *e outras histórias* insere-se modestamente nos festejos desta data. É, na verdade, do ponto de vista dos editores, uma dupla homenagem: a Cesar Lattes, que tanto fez pela Ciência no Mundo e no Brasil, e a Alfredo Marques, que nos deixou há menos de 2 anos, grande amigo nosso e de Lattes. Sem seu estímulo e colaboração não teríamos feito a primeira edição.

Alguns comentários relacionados a decisões editoriais se fazem necessários. Em primeiro lugar, os originais estavam todos em Word e foram redigitalizados em LaTeX e revisados. Nesta árdua tarefa, contamos com a inestimável colaboração de Francisca Valéria Fortaleza de Vasconcelos, a quem somos muito gratos. Procuramos, dentro do possível, padronizar o formato dos diversos artigos. Aqueles em língua portuguesa tiveram a ortografia revista conforme as novas regras ortográficas do Acordo da Língua Portuguesa. Além disso, pequenos erros de digitação e alguns lapsos da primeira edição foram corrigidos pelos editores. As traduções que acompanhavam os artigos de Wataghin e Occhialini foram suprimidas dessa edição, por questão de espaço. Quando oportuno, algumas poucas notas dos editores (N.E.) foram incluídas em rodapé em certos textos, esclarecendo ou detalhando alguma informação. Uma última observação cabe registrar: refere-se à grafia do nome de Lattes. Todos sabem que seu nome de batismo é *Cesare*, mas quase sempre se escreve *César* ou *Cesar*. Consultando como o próprio Lattes assinava as atas do CBPF, optamos por usar nesse livro sempre a forma sem acento.

Por fim, agradecemos a Felipe Silveira, pelo auxílio no preparo de algumas figuras e na revisão final do manuscrito, e ao editor da LF, José Roberto Marinho, por ter aceitado publicar a presente obra.

2

Prefácio

José Israel Vargas [1]

Evocar o momento científico da descoberta do méson-π, apresentar suas múltiplas projeções, reconstruir o clima em que transcorreu, revisitar seus desdobramentos, eis a proposta deste livro.

À distância (o intervalo aqui já venceu os cinquenta anos), a restauração dos episódios torna-se mais fácil pela melhor definição histórica do contexto, pela nitidez com que se tornam visíveis os caminhos que vieram a ser consolidados em rotas permanentes. As trilhas alternativas, aquelas que o tempo sepultou junto às expectativas, aspirações, empenhos e canseiras dos desbravadores, estas desaparecem sob os próprios escombros. Não se afigura, portanto, simples, a tarefa de restaurar o clima vivido, sobretudo quando se pretende caminhar por todos os planos de sua construção, em particular aqueles que se ergueram no talento, na fibra, enfim, no valor humano de seus protagonistas.

Foi essa a delicada e difícil tarefa que se propuseram os organizadores deste volume, convictos da importância de que o tema se reveste, pelas implicações no desenvolvimento da física do século XX e pelas marcadas projeções no desenvolvimento científico brasileiro.

Em seu benefício contaram com a disponibilidade de preciosa documentação referente, na maior parte, à contribuição de *Cesar Lattes* aos memoráveis feitos

---

[1] *N.E.*: O Prof. José Israel Vargas, Ministro de Ciência e Tecnologia do Governo brasileiro no período 1994-1998, é Presidente da Academia de Ciências do Terceiro Mundo. Doutor pela Universidade de Cambridge, foi membro eleito durante sete anos e meio e presidente do Conselho Executivo da UNESCO (1987-89). Professor Emérito da Universidade de Minas Gerais e do Centro Brasileiro de Pesquisas Físicas, contemporâneo dos eventos relatados neste livro e tendo sua trajetória científica fortemente influenciada por eles, participou ativamente de muitos e de seus desdobramentos.

cercando aquela descoberta, acumulada graças ao zelo de muitos, sensíveis à histórica relevância. Contaram, sobretudo, com o testemunho de protagonistas daqueles episódios. Relevam, nesse particular, o próprio testemunho de *Cesar Lattes*, de alguns de seus colegas em Bristol, onde tiveram lugar os primeiros passos daquela fascinante aventura, e no Imperial College de Londres, onde um projeto semelhante teve curso simultânea e independentemente.

Os desdobramentos dessa descoberta, suas implicações na modelagem da nova física da segunda metade deste século, os impulsos que vieram renovar o ambiente científico brasileiro, abrindo espaços novos na pesquisa – como, por exemplo, a criação do CBPF, do IMPA e do INPA – além das fronteiras da medicina tropical, onde já havíamos alcançado maioridade, e, em âmbito latino-americano, a criação, do Laboratório de Física Cósmica de Chacaltaya, são destaques no elenco de relatos. Não faltaram também referências às instituições que se criaram ao ensejo daquela descoberta, vindo a desfrutar de uma posição destacada em todo o desenvolvimento científico posterior, como foi o caso do CNPq, e aos nomes de tantos que militaram nessas lides e contribuíram com esforço e talento para torná-las realidade.

O instante em que este documento é lançado marca o regozijo pelo êxito científico da descoberta do méson-π e pelos novos caminhos que viabilizou, agora trilhados pelas gerações que buscam realização científica. Muito em particular, é um testemunho da gratidão a *Cesar Lattes*, cujo talento científico e amor à sua terra e gente foram decisivos para ensejar o crescimento e a diversificação da pesquisa científica brasileira até os dias correntes. É imenso o prazer que venho desfrutando, ao longo dos anos, honrado com a calorosa amizade deste insigne cientista.

3

Nota dos Editores

Francisco Caruso, Alfredo Marques & Amós Troper

Nosso objetivo principal com a edição deste livro é rememorar os 50 anos da descoberta do méson-π, pelo grupo de raios cósmicos de Bristol, que contou com a decisiva participação do cientista brasileiro *Cesar Lattes*.

O livro, escrito de forma não convencional (e por muitas mãos), apresenta, nos seus diversos artigos, aspectos vários dessa descoberta científica, bem como de seus inegáveis reflexos no Brasil. De fato, o enorme prestígio científico internacional adquirido pelo jovem cientista *Lattes* (com 24 anos em 1948) serviu de alicerce para a fundação do Centro Brasileiro de Pesquisas Físicas, em 1949, e foi decisivo na criação do Conselho Nacional de Pesquisas (CNPq), em 1951, entre outros fatos relevantes.

O impacto científico da descoberta do píon no cenário da Física Contemporânea pode ser aferido, por exemplo, do elenco das grandes descobertas da Física, de 1815 até 1984, feito pelo eminente historiador da Física Moderna, Abraham Pais, em seu livro *Inward Bounds*. Em dois anos consecutivos estão presentes as duas famosas descobertas envolvendo o nome de *Cesar Lattes*, a saber: "*1947: – Descoberta experimental de uma segunda radiação cósmica mesônica, logo denominada píon (...); 1948: – Os primeiros píons artificialmente produzidos detectados em Berkeley (...)*".

E como aferir a importância da participação de Lattes nessas descobertas? Tentativas de obter dele as respostas conduzem invariavelmente ao mesmo ponto. Em recente entrevista sobre este assunto, Lattes comenta a sua participação científica nos eventos acima mencionados, com a modéstia que lhe é peculiar, limitando-se a dizer: "*Fui empurrado para a história*", e arremata sem alterar o estilo sóbrio de seu depoimento: "*... e fiz o possível*". Entretanto, durante sua última visita ao CBPF, em setembro de 1998, quando estávamos concluindo a revisão final do texto e escrevendo esta nota introdutória, Lattes surpreendeu-nos com um depoimento único, como ele próprio confirmou depois. Deixando a modéstia de lado por poucos segundos, mas sem abandonar seu refinadíssimo senso de humor, ao ser apresentado por um de nós a um grupo de professoras que visitava o CBPF como "o famoso professor Lattes", ele disse: – "*da fama eu não gosto; só aprecio a glória...*", e começou a rir. Esta frase nos fez lembrar de Borges, para quem a glória é uma forma de incompreensão (talvez a pior). Percebemos então, naquele momento, com maior clareza, que o que pretendemos ao coligir os textos neste volume é contribuir para a difusão de um entendimento mais profundo – e consequentemente "menos glorioso" – deste momento tão marcante da obra científica de Cesar Lattes: a descoberta do méson-π. Através desta compreensão, despojada de qualquer mistificação, talvez o significado desta descoberta para a mudança de cenário da Ciência brasileira seja melhor percebido, para a glória de nossa Cultura.

Em suma, ainda que de modo imperfeito, este livro evoca não somente a descoberta do méson-π e as novas fronteiras da Física Moderna por ela abertas, bem como alguns aspectos relevantes na criação da moderna ciência no Brasil, enfatizando, particularmente, os primeiros anos que se seguiram ainda sob o forte impacto do trabalho de Cesar Lattes.

Os artigos foram escritos por diferentes colaboradores diversamente relacionados a Lattes, à descoberta e à produção do méson-π. Procuramos caracterizar essa condição agrupando-os da seguinte forma: depoimentos de precursores, testemunhos de participantes, "relatos da posteridade" e, finalmente, aqueles que focalizam desdobramentos diversos daquelas descobertas.

Como o leitor logo perceberá, há, na maioria dos artigos, longas referências a diferentes aspectos daqueles feitos (algumas, de certa forma, repetitivas), de modo que o conjunto mostra um certo conteúdo redundante. Esta foi, entretanto, uma escolha deliberada; consideramos aqui a redundância construtiva, pois acreditamos que através dela seja mais fácil destacar dos relatos seu conteúdo consolidado, sem, no entanto, apagar as nuances de opiniões ou até mesmo afastar os pontos de polêmica. Deixamos ao leitor a difícil tarefa de ponderar o que não é consenso.

Cabe agora um breve comentário sobre alguns critérios editoriais por nós adotados. Além da natural revisão dos textos em seus aspectos formais – de ortografia e tipografia –, em um número muito limitado de casos, em que a redação original omitia uma palavra de significado óbvio, entretanto essencial, os editores tomaram a liberdade de completar o texto. Essa decisão poupou o dispêndio de longo tempo na troca de correspondência com os autores por motivos tão triviais, as quais, no entanto, poderiam atrasar em muito o lançamento do livro. Em outros casos, pequenos reparos gramaticais óbvios também foram aplicados e, sempre que possível, os autores foram consultados por telefone. Acreditamos ter interpretado corretamente o curso das ideias expostas pelos autores e em nenhum caso praticado intervenção que as deformasse. De qualquer forma, desde já, nos desculpamos com os autores por eventuais erros de revisão que certamente ainda nos escaparam. No restante, cabe notar que os artigos aqui reproduzidos não necessariamente expressam as opiniões dos editores.

Nossos primeiros agradecimentos vão aos colaboradores deste livro que com a força de suas experiências, de suas memórias e de suas penas tornaram esta edição possível. Agradecemos também a Márcia de Oliveira Reis Brandão, María Elena Pol e Susana Zanetti de Caride pelo auxílio inestimável na revisão final dos textos; a Cesar Lattes, a Ana Maria Ribeiro de Andrade, a Cássio Leite Vieira, a Edison Shibuya e ao CBPF por terem colocado a nossa disposição seus arquivos iconográficos e ao CNPq pelo indispensável apoio dado à publicação deste livro. Aos Profs. Ubyrajara Alves e Evandro Mirra de Paula e Silva pelo apoio dado a esta edição.

Por último, devemos confessar que o ato de ler e reler estes textos muitas vezes, nos últimos meses, fez com que, inconscientemente, buscássemos adjetivar Cesar Lattes. Lembramo-nos, então, do curioso sentimento de posse de Otto Maria Carpeaux, com relação a Dante Alighieri, expresso no título de seu belo artigo *"Meu Dante"*. Gostaríamos, portanto, que nos fosse permitido usar o mesmo recurso e que o *nosso* Prof. Cesar Lattes aceitasse este volume como demonstração de nosso enorme afeto e apreço por ele.

4

Saudações a Cesar Lattes por ocasião de seus 60 anos

Gleb Wataghin[1]

Festejamos neste 1984 os 60 anos de Cesar Lattes e também faz cinquenta anos desde que, no já longínquo 1934, cheguei a São Paulo e iniciei a minha atividade de docente na Faculdade de Física. Estas duas recorrências estão associadas para mim a um significado que vai além da coincidência, porque de todos os estudantes que lá sucederam a Cesar Lattes, ele é aquele que melhor do que qualquer outro respondeu aos meus esforços de professor, e não creio, ao dizer isto, estar cometendo uma injustiça a qualquer um de tantos galhardos estudiosos que foram meus alunos. Os resultados científicos alcançados por Cesar lattes com a descoberta e produção artificial do méson-π são tais que o colocam no primeiro plano da Física brasileira, nos primeiríssimos do mundo, e me deixam orgulhoso de poder dizer que foi meu aluno. Mas além do respeito e da estima pelo talento do estudioso, uma profunda amizade me liga a Cesar Lattes, cujas bases foram estabelecidas nos anos em que o jovem Lattes afrontava, cheio de entusiasmo, as suas primeiras provas como cientista e que depois se desenvolveu no curso de uma fecunda colaboração no estudo dos raios cósmicos e, in particular, da radiação penetrante. Esta amizade não se interrompeu após o meu retorno à Itália e foi para mim uma alegria revivê-la quando, por iniciativa do próprio Lattes, reencontrei, já aposentado da Universidade italiana, o caminho do Brasil e, se assim posso dizer, um prolongamento de juventude na Universidade de Campinas. É, portanto, para mim particularmente agradável me associar a esta celebração, com as mais

---

[1] *N.E.*: Texto publicado originalmente em italiano como introdução de *Topics on Cosmic Rays: 60th anniversary of C.M.G. Lattes*, vol. 1, Campinas, Editora da UNICAMP, 1984, por ocasião dos 60 anos de Cesar Lattes. Gleb Wataghin faleceu em 1986. Apresenta-se, aqui, apenas a tradução para o português feita por F. Caruso e A. Marques. A versão inglesa e a original foram omitidas nessa segunda edição.

calorosas congratulações ao amigo Lattes e os mais fervorosos votos de longos e fecundos anos de trabalho científico.

5

Cesar Lattes: Bristol Years

Giuseppe Occhialini

> "To see infinity in a grain of sand and eternity in an hour"

When choosing that title I had in mind to reproduce a retrospective sight of what was the 4th floor of the Royal Fort, the "cigarette tower" where the works of the H.H. Wills Physical Laboratory took place. I watched it to evolve from a few pieces of dusty equipment, its quietness during the war, towards an international research centre, with plenty of enthusiasm, frenetic very often, of young physicists. Along those evolution years I have seen Cesar Lattes, first one of a young generation marked by intense dedication to research work.

Next I have checked upon dates. I have found that a little over two years had gone by. However I could yet to write about them as if they took as long as an era, so many were the significant events that have happened there. Some relativists use to say that time is measured by events but do not frame them. In that sense I may estimate Lattes' stay in Bristol as lasting at least one decade.

It was on January 1946. Feeling the enormous possibilities of the new concentrated nuclear emulsions, I suggested Powell to invite Lattes to Bristol.

We felt that technical progress was favourable to new discoveries and that previous knowledge presented many flaws. We needed new, younger efforts. At that time young British physicists reassumed slowly their positions at the universities, supported by "lavish" grants offered by government. However most of them did not feel attracted by that second class technique, "routine of little practical use"

(Smythe report on atomic weapons) in a countryside university. Physicists in occupied Europe slowly reconstructed their fragments of life. Then I remembered Lattes asking me, when I left Brazil, to call him up just in case some research opportunity could be opened. By doing so I felt to be paying part of my debt to Brazil.

Lattes arrived and the life in the 4th floor of Royal Fort changed since then. He brought in the restlessness of springtime and the exuberance of his young energy to that atmosphere of sober and determined dedication. Generous, he invited his friend Ugo Camerini who arrived with twenty social shirts packed up by his mother (which we shared on due time) to begin a career extremely brilliant, witness of the vitality of the young school of the University of São Paulo.

This way started what was, for the lower premises in the "cigarette tower" the exotic life of the 4th floor. Unshaved, sometimes unwashed, working seven days a week until two, sometimes four o'clock in the morning, drinking rather strong coffee, running up and down, shouting, quarrelling and laughing we were taken with sympathy by all inmates, tired of the war, and by the foreigners to Royal Fort.

After a long war they tasted the delights of peace, order and safety, and the dreams of an undisturbed academic future.

Those in the 4th floor had not experienced the problems of war, but carried with them many complexes and frustrations seriously prepared to grab the chance to show their character in a civil scientific job.

Outside the laboratory the feelings of population were of a mixed kind. Behind the Royal Fort the Robin Hood pub accepted gladly our invasions, five minutes before closing time, to drink a pint of beer before going back to work. After midnight Lattes apportioned horrible cigarettes made after old tips gathered all over the Institute. The local priest included a reference to us in a sermon: "They work hard but have no guidance". In doubt, a policeman followed the steps of Lattes and Camerini to their home, after spotting them walking astray along the calm streets of Bristol, four o'clock in the morning, talking a strange idiom and suspiciously sobers.

That was the external view. The inner reality was not as romantic but anyhow more exciting. It was the reality of hard work, intense and continuous – of deep excitation and unbelievable dreams that came through, at the end. It was the reality of discovery and the role of Lattes was remarkable. When he arrived at Bristol the activity in the 4th floor was restricted to nuclear physics. He worked on the disintegration of Samarium and in experiments of nuclear scattering proposed by Powell but his heart moved him out to cosmic rays. Meanwhile he dominated the new technique. With Peter Fowler, then a post graduate student, he established the range-energy relations in the new concentrated emulsions.

With Camerini he investigated the fading of latent images. After their discovery that even in the new emulsions the gradual fading of the images was important, it became clear that the exposure time of the plates that I had brought to the Pic du Midi had to be shortened.

The gain of a few weeks in exposure time owing to those findings revealed to be of utmost value. At the time other groups exposed plates in airplane flights and we, without realizing it, took part in the race to find the pion-muon decay.

When he saw the first plates exposed to cosmic rays, Lattes went very much excited. All his studies and experience in cosmic rays at São Paulo concentrated on those new evidences. He recognized immediately the first disintegration of the meson that I showed him. After that moment he asked me, literally required, to take place in that new venture.

His contribution was quite important. He brought in not only his ambition and young energy but also physical insight, neatness of thought and long and passionate studies of cosmic ray physics. In São Paulo he was criticized as a man of much thought and little action. In Bristol he exploited in action every fruit of his studies. An example will show what I mean. Before the discovery of pion-muon decay, Lattes and Camerini performed a precise analysis of the products of meson disintegration and showed that the total energy produced was greater than the mass equivalent accepted for mesotrons. That was a task proper of an authentic physicist.

The discovery and analysis of the pion-muon decay is part of history of science. The observation of the artificial production of pions when Lattes left Bristol and moved on to Berkeley belongs to another epoch. If he had only worked those two years in Bristol, he deserved already a place in that history.

But history is an academic and abstract concept. More immediate should be the widespread consensus of the younger and weaker then him, which he has always defended. His "quixotic" character brought him into trouble and difficulty many times. Dom Quijote is a hero of rationalists but, as a character, he was and is even today recalled with love and appraisal.

To Cesar Lattes of nowadays, of my past memories, this is probably my best tribute. To be a great physicist is rather difficult, but mean characters can also be such. Nobility of character is innate.

6

My Work in Meson Physics with Nuclear Emulsions

Cesar Lattes[1]

At the end of the Second World War, I was working at the University of São Paulo, Brazil, with a slow meson triggered cloud chamber, which I had built in collaboration with Ugo Camerini and A. Wataghin. I sent pictures obtained with this cloud chamber to Giuseppe P.S. Occhialini, who had recently left Brazil and had joined Cecil F. Powell at Bristol. Upon receiving from Occhialini positive prints of photomicrographs of tracks of protons and α-particles, obtained in a new concentrated emulsion just produced experimentally by Ilford Ltd., I immediately wrote to him asking to work with the new plates, which obviously opened great possibilities. Occhialini and Powell arranged for a grant from the University of Bristol; I somehow managed to get to Bristol during the winter of 1946.

I was given the task of obtaining the shrinkage factor of the new emulsion (which was much more concentrated than the old ones); Occhialini and Powell were still at work on n-p scattering at around 10 MeV, using the old emulsions.

I decided that the time allotted to me at the Cambridge Cockroft-Walton accelerator, which provided artificial disintegration particles as probes for the shrinkage factor, was sufficient for a study of the reactions:

$$D(d,p)H_1^3 \qquad Be_4^9(d,p2n)Be_4^8$$
$$Li_3^6(d,p)Li_3^7 \qquad B_5^{10}(d,p)B_5^{11}$$
$$Li_3^7(d,p)Li_3^8 \qquad B_5^{11}(d,p)B_5^{12}$$

---

[1] Reprinted from *Proceedings of the Symposium on the History of Particle Physics* – The Birth of Particle Physics, Fermilab, May 1960.

Through analysis of the tracks, we obtained a range-energy relation for protons up to about 10 MeV, which was used for several years in research where single charged particles were detected, e.g., pions and muons [1].

In the same experiment, I placed borax-loaded plates, which Ilford had prepared, at my request, in the direction of the beam of neutrons from the reaction

$$B_5^{11} + H_1^2 \Rightarrow C_6^{12} + n_o,$$

which gives a peak of neutrons at about 13 MeV. The idea, which worked well, was to obtain the energy and momentum of neutrons, irrespective of their direction of arrival (which was not known), through the reaction

$$n_o^1 + B_5^{10} \Rightarrow He_2^4 + He_2^4 + H_1^3$$

I asked Occhialini, who had decided to take a vacation in the Pyrenees (Pic-du-Midi and environs), to take with him for about a one month exposure, boxes of emulsions; some were loaded with borax, some were normal plates (without borax). All were made of the new concentrated B1 type emulsion for which a range-energy relation already existed. The normal plates were to be used for the study of low energy cosmic rays and as a control, to see if we were detecting cosmic ray neutrons.

When Occhialini processed the emulsions, on the same night in which he arrived back from vacation, it became clear that borax loaded emulsions had many more events than the unloaded ones; borax somehow kept the latent image from fading; normal plates had a great amount of fading. The variety of events in the borax plates, and the richness in detail, made it obvious that the neutron energy detection was but a side result. The normal events seen in the plates were such as to justify putting the full force of the Laboratory into the study of normal low-energy cosmic ray events. After a few days of scanning, a young lady – Marietta Kurz – found an unusual event:[2] one stopping meson and, emerging from its end, a new meson of about 600 μ range, all contained in the emulsion. I should add that mesons are easily distinguished from protons in the emulsion we used, due to their much larger scattering and their variation of grain density with range. A few days later a second 'double' meson was found; unfortunately in this case the secondary did not stop in the emulsion but one could guess, by studying its ionization (grain counting), that its extrapolated range was also $\sim 600\,\mu$. The first results on the 'double' meson were published in *Nature*[2] By the way, the cosmic ray neutrons (direction, energy) were also obtained in the same plates and the results published in the same volume of *Nature*[3].

Having one and a half 'double mesons' which seemed to correspond to a fundamental process (although it could have been an exothermal reaction of the

---

[2] *N.E.*: The sequence in which the two first mesons were observed was mistaken in the text; the one with a complete secondary within the emulsion's thickness was actually the second to be found.

type $\mu^- + X_a^b \Rightarrow X_a^{b-1} + \mu^+$), the Bristol group realized that one should quickly get more events. I went to the Department of Geography of Bristol University and found that there was a meteorological station at about 18 600-ft, about 20 km by road from the capital of Bolivia, La Paz. I therefore proposed to Powell and Occhialini that if they could get funds for me to fly to South America, I could take care of exposing borax loaded plates at Chacaltaya Mountain for one month. That was done and I left Bristol with several borax-loaded plates plus a pile of pound notes sufficient to carry me to Rio de Janeiro and back. Contrary to the recommendation of Professor Tyndall, Director of the H.H. Wills Laboratory, I took a Brazilian airplane, which was wise, since the British plane crashed in Dakar and killed all its passengers.

After the agreed time I developed one plate in La Paz. The water was not appropriate and the emulsion turned out stained. Even so it was possible to find a complete 'double meson' in this plate; the range of the secondary was also around 600 μ!

Back in Bristol the plates were duly processed and scanned; about 30 'double mesons' were found. It was decided that I should try to get the mass ratio of the first and second mesons by doing repeated grain counting on the tracks. The results convinced us that we were dealing with a fundamental process [4]. We identified the heavier meson with the Yukawa particle and its secondary with Carl Anderson's mesotron. A neutral particle of small mass was needed to balance the momenta.

At the end of 1947 I left Bristol with a Rockefeller Scholarship with the intention of trying to detect artificially produced pions at the 184" cyclotron which had started operation at Berkeley, California. The beam of α-particles was only 380 MeV (95 MeV/nucleon), an energy insufficient to produce pions. I took my chance on the 'favourable' collisions in which the internal momentum of a nucleon in the α and the momentum of the beam provided sufficient energy in the centre of mass system. Results showed that mesons were indeed being produced. Two papers describe the method of detection and the results, the first referring to negative pions, the second to positive [5]. By making use of the range of pions and their curvature in a magnetic field, it was possible to estimate the masses to be about 300 electron masses.

Around February 1949, I was preparing to leave Berkeley to return to Brazil. At that time Edwin McMillan, who had his 300 MeV electron synchrotron in operation, asked me to look at some plates which had been exposed to γ-rays from his machine. In one night I found about a dozen pions, both positive and negative and the next morning delivered to McMillan the plates and maps which allowed the finding of the events. I do not know what use McMillan made of the information, but there is no doubt that they were the first artificially photoproduced pions detected.

Bibliography

[1] C.M.G. Lattes, P.H. Fowler, and P. Cuer, "Range-Energy Relation for Protons and α-Particles in the New Ilford 'Nuclear Research' Emulsions", *Nature* **159** (1947), 301-2; C.M.G. Lattes, P.H. Fowler and P. Cuer, "A Study of Nuclear Transmutations of Light Elements by the Photographic Method", *Proc. Phys. Soc. (London)* **59** (1947), 883-900.

[2] C.M.G. Lattes, H. Muirhead, G.P.S. Occhialini, C.F. Powell, "Processes Involving Charged Mesons", *Nature* **159** (1947), 694-7.

[3] C.M.G. Lattes and G.P.S. Occhialini, "Determination of the Energy and Momentum of Fast Neutrons in Cosmic Rays", *Nature* **159** (1947), 331-2.

[4] C.M.G. Lattes, G.P.S. Occhialini and C.F. Powell, "Observations on the Tracks of Slow Mesons in Photographic Emulsions", *Nature* **160** (1947), 453-6 and 486-92; C.M.G. Lattes, G.P.S. Occhialini and C.F. Powell, "A Determination of the Ratio of the Masses of π^- and μ^- Mesons by the Method of Grain-Counting", *Proc. Phys. Soc. (London)* **61** (1948), 173-83.

[5] Eugene Gardner and C.M.G. Lattes, "Production of Mesons by the 184-Inch Berkeley *Cyclotron*", *Science* **107** (1948), 270-1; John Burfening, Eugene Gardner and C.M.G. Lattes, "Positive Mesons Produced by the 184-Inch Berkeley Cyclotron", *Phys. Rev.* **75** (1949), 382-7.

7

The Development and use of Nuclear Emulsions in England in the Years 1945-50

G.D. Rochester[1]

Introduction

I would like to add a little to the part of Professor Rossi's lecture concerned with the development and use of nuclear emulsions in England in the years 1945-50. Much had been achieved before 1945, especially by Powell and his collaborators, but the emulsions then available were severely limited by their lack of sensitivity.

Emulsion Development 1945-6

The first major step in the improvement of sensitivity was taken almost at the same time, and certainly independently, in the period 1945-6, by a French-Canadian chemist, Demers, and a research team, led by Waller, at Ilford Ltd., in England, by the achievement of a large increase in the silver bromide: gelatine ratio without loss of chemical stability or increase in the background. Demers made

---

[1] *N.E.*: Reprinted from the *Proceedings of Symposium on the History of Particle Physics – The Birth of Particle Physics*, Fermilab May 1980. Prof. Rochester sent for publication this written version of the comment made after Prof. B. Rossis's talk.

his technical advance in the second half of 1945 as part of an attempt to perfect a new technique in nuclear research; Ilford made their advance as a direct result of national planning. The person responsible for the executive side of the planning was undoubtedly Blackett, who under the Labour Government, elected on July 1945, became the chairman of a powerful scientific committee of the Ministry of Supply. One of the tasks assigned to this committee was to consider the future of nuclear research outside the defence establishments and, in furtherance of the decision to encourage such research, it set up two panels, one concerned with the development of cyclotrons, and other, nuclear emulsions. It seems probable that Chadwick, who was familiar with American and Canadian work in these fields, must have advised Blackett and, indeed, the chairman of both committees was Rotblat, a member of Chadwick's staff at Liverpool University. It is also possible that Blackett's decisions were influenced not only by his high regard for Powell's work but by his friendship with Occhialini, who went to Bristol in 1945 on Blackett's recommendation.

The first meeting of the Emulsion Panel was held on 21st November, 1945, under the chairmanship of Rotblat, with representatives of Ilford Ltd., and various universities. The university representatives were Livesey, (Cavendish Laboratory, Cambridge University), May, (King's College, London University), and Powell, (Bristol University). Occhialini was not a member of the Panel but he had a considerable influence on its work. I became a member of the Panel later. Funds were made available by the Ministry of Supply and the main function of the Panel was to give contracts in those areas considered to be of primary scientific interest and importance. From the outset the main aim was seen to be the production of emulsions capable of recording charged particles of relativistic velocities, *i.e.*, sensitive to minimum ionisation. Secondary aims were the development of suitable microscopic equipment, the loading of emulsions with various elements, and test facilities.

By the third meeting of the Panel, held on 7th May 1946, Ilford had produced beautiful new emulsions in four grain sizes, A, B, C, and D, and three sensitivities, 1, 2 and 3. The silver bromide : gelatine ratio had been increased to eight times above normal emulsion. As an example of sensitivity and grain size it may be noted that the C2 emulsion, later much used in cosmic ray research, had a sensitivity of six times minimum and a grain size of 0.15 μm. At one of the early meetings a Demers' plate sent by Chadwick was shown, and I recall that its sensitivity was somewhat higher than the best Ilford emulsion and that the grain size was only 0.08 μm. Later, the Demers' formula was made available to English manufacturers, but it seems to have had little effect because the Canadian's method were unsuited to manufacture on a commercial scale.

The production of better emulsions immediately led to their use in cosmic ray research. Exposure of plates by Perkins (B1) and by Occhialini and Powell (C2) in 1946, led to the discovery of the π-meson.

Electron Sensitive Emulsions

For the first year of the Panel's existence contracts were given only to Ilford Ltd., but in 1947 English Kodak Ltd., was invited to participate. By early 1948 the chief emulsion chemist of Kodak, Berriman, and his staff, had produced much improved emulsions and by November of the same year a sample of the first emulsion sensitive to minimum ionisation (the NT4) was presented to the Panel. Ilford followed in September 1949 with the G5 emulsion and this became the one mainly used in cosmic ray research.

Comments

It is remarkable how the right technique emerged at the right time through the combined effort of scientist and manufacturer and that a medium was produced which rivalled the cloud chamber in the visual beauty of its recording of transient 'events'. As is well known the realisation of the immense scientific potential of the emulsion was due almost entirely to Powell and the Bristol school.

What has been stressed in this note is the part played by technical advances, especially the English manufacturers' contributions to the production of electron sensitive emulsions. To this could be added the later production by the same manufacturers of thick emulsions, of blocks of emulsions, and the concomitant technique of the 'temperature cycle method' invented by Dilworth, Occhialini and Payne of Bristol. Again, the fact that an emulsion like G5 had a relatively large grain size of 0.3 μm allowed Powell to introduce mass scanning by a large team of ladies (sometimes facetiously called "Cecil's Beauty Chorus"!) with low power microscopes, and thus to obtain relatively quickly a wealth of data for his large team of scientists. It is unlikely that mass scanning could have been employed with finer grained emulsions such as, for instance, the Demers' emulsions.

8

Cesar Lattes Half a Century Ago: The Pion Pioneers

Owen Lock[1]

While the classic discoveries of Thomson and Rutherford opened successive doors to subatomic and nuclear physics, particle physics may be said to have started with the discovery of the positron in cosmic rays by Carl Anderson at Pasadena in 1932, verifying Paul Dirac's almost simultaneous prediction of its existence.

Anderson used a cloud chamber, expanded at random, in a high magnetic field. At the same time, Patrick Blackett at Cambridge was joined by an inventive Italian, Giuseppe Occhialini, sent by a master of counter coincidence techniques, Bruno Rossi, then in Florence, to learn about cloud chambers. Very soon Blackett and Occhialini had built a counter-controlled chamber with which they discovered electron-positron pair production, a key prediction of Dirac's ideas.

Cloud chambers played a major role in cosmic ray studies in the following years, leading to the discovery of the 'mesotron' in 1937, originally identified as the nuclear force carrier postulated by Hideki Yukawa in 1935. However, several difficulties soon arose with this hypothesis, even though pictures of its decay to an electron, as postulated by Yukawa to explain beta-decay, were observed in cloud chamber pictures in 1940. In particular, the mesotron appeared to have a very weak nuclear interaction with matter, conclusively demonstrated by the counter experiments of Marcello Conversi, Ettore Pancini and Oreste Piccioni in Rome from 1943-1947.

---

[1] N.E.: Reprinted from "Physics Monitor", *CERN Courier*, June 1997, pp. 2-6, without pictures.

A possible explanation of these difficulties had been put forward in Japan in 1942 and 1943 by Yasutaka Tanikawa and by Shoichi Sakata and Takeshi Inoue, who suggested a two-meson hypothesis with a Yukawa-type meson decaying to a weakly interacting mesotron. Because of the war their ideas were not published in English until 1946 and 1947, the journals in question not reaching the USA until the end of 1947.

Unaware of the Japanese work, Robert Marshak had put forward a similar hypothesis in June 1947, at a conference of American theoreticians on Shelter Island (off Long Island), and which he published later that year with Hans Bethe. None of the scientists at the conference knew that such two-meson decay events had already been observed some weeks earlier by Cecil Powell and his collaborators in Bristol, using the then little known photographic emulsion technique, but which in Powell's hands became a powerful research tool.

Powell had been a research student under C.T.R. Wilson at the Cavendish Laboratory in Cambridge, before joining the H.H. Wills Physics Laboratory (also known as the Royal Fort), at Bristol University in 1928 as an assistant to the Director, Arthur Tyndall. They worked together on the mobility of ions in gases until 1935 when Powell became interested in nuclear physics, inspired by the discoveries in Rutherford's Cavendish Laboratory. Together with a young lecturer, Geoffrey Fertel, he embarked on the construction of a 750 keV Cockcroft-Walton accelerator, which they brought into operation in 1939.

The original intention was to study low energy neutron scattering using a Wilson cloud chamber. However, in 1938 the theoretician Walter Heitler (then in Bristol) mentioned to Powell that in 1937 two Viennese physicists, Marietta Blau and Herta Wambacher, had exposed photographic emulsions for five months at 2 300 m in the Austrian Alps and had seen the tracks of low energy protons as well as 'stars' or nuclear disintegrations, probably caused by cosmic rays. Heitler commented that the method was so simple that 'even a theoretician might be able also to do it'. This intrigued Powell, and he and Heitler travelled to Switzerland with a batch of Ilford half-tone emulsions, 70 microns thick, and exposed them on the Jungfraujoch at 3 500 m. In a letter to *Nature* in August 1939, they were able to confirm the observations of Blau and Wambacher.

The half-tone emulsions could only record the tracks of low energy protons and alpha particles and Powell realised that to do useful work it was necessary to increase their sensitivity by increasing the concentration of silver bromide.

World War II interrupted the work, but with the existing emulsions Powell was able to show that for scattering studies they gave results superior to cloud chambers, as well as being much faster.

Blackett (who had been a contemporary of Powell in the Cavendish Laboratory) then played a decisive role through his influence with the Ministry of Supply of the

1945 UK Labour Government. He was largely responsible for the setting up of two panels, one to plan accelerator building in the United Kingdom (which he chaired) and one to encourage the development of sensitive emulsions (chaired by Joseph Rotblat, recently awarded the Nobel Peace Prize for his Pugwash work).

Towards the end of the war, Blackett had invited his erstwhile collaborator Occhialini, then in Brazil, to join the British team working with the Americans on the atomic bomb. Occhialini arrived in the United Kingdom in mid-1945, only to learn that, as a foreign national, he could no longer work on the project. Instead, he joined Powell in Bristol, becoming a driving force behind the development of the new emulsion technique. He was joined by one of his research students, Cesare Lattes, toward the end of 1946.

Photographic manufacturers Ilford were soon able to supply 'Nuclear Research Emulsions' and in autumn 1946 Donald Perkins, then at Imperial College, London, exposed at 9 100 m in a RAF aeroplane, while Occhialini took several dozen plates to the Pic du Midi at 2 867 m in the French Pyrenees. At that time access to the Pic was by a rough track in summer and by ski in winter, a small téléphérique only being brought into service in the summer of 1952, but Occhialini had been a mountain guide in Brazil.

Examination of the emulsions in Bristol and in London revealed, as Powell latter wrote, "a whole new world. It was as if, suddenly, we had broken into a walled orchard, where protected trees flourished and all kind of exotic fruits had ripened in great profusion". This new world became a subject of intensive investigation. Occhialini has well described the atmosphere at Bristol: – "Unshaved, sometimes I fear unwashed, working seven days on the week till two, sometimes four in the morning, brewing inordinately strong coffee at all hours, running, shouting, quarrelling and laughing, we were watched with humorous sympathy by the war-worm native denizens of the Royal Fort" ⋯ "It was really of intense, arduous and continuous work, of deep excitement and incredibly fulfilled dreams. It was the reality of discovery⋯".

Perkins was the first to observe a clear example of what appeared to be the nuclear capture of a meson in the emulsion and producing a nuclear disintegration. Measurements of the multiple scattering as a function of residual range indicated a mass between 100 and 300 times that of the electron. Perkin's observations, published in January 1947, were confirmed by Occhialini and Powell, who published details of six such events only two weeks later. Mesons were easily distinguished from protons in the emulsion because of their much larger scattering and by their variation of grain density with range.

Yet another exotic fruit followed. In the spring of 1947 one of Powell's team of microscope observers, Marietta Kurz, found a meson stopping and giving rise to a second meson, which left the emulsion when nearly at the end of its range. Powell and a young Bristol graduate, Hugh Muirhead, were the first physicists to look at the event, which they immediately recognised as being two related mesons. Within a few days a similar event was found by Irene Roberts, the wife of the group technician, Max Roberts, who later worked at CERN for many years. In this event the secondary meson ended in the emulsion, with a range of 610 microns.

The two events gave such convincing evidence for a two-meson decay chain that Lattes, Muirhead, Occhialini and Powell published their findings in '*Nature*' in the issue of 24 May, 1947. Commenting on the problems surrounding the identification of the cosmic ray mesotron with the Yukawa nuclear force meson, they wrote: "Since our observations indicate a new mode of decay of mesons, it is possible that they may contribute to a solution of these difficulties".

More evidence was needed to justify such a radical conclusion. For some time no more two-meson events were found in the Pic du Midi emulsions and it was decided to make exposures at much higher altitudes. Lattes proposed going to Mount Chacaltaya in the Bolivian Andes, near the capital La Paz, where there was a meteorological station at 5 200 m. Arthur Tyndall recommended that Lattes should fly BOAC to Rio de Janeiro. Lattes preferred take the Brazilian airline Varig, which had a new plane, the Super Constellation, thereby avoiding a disaster when the British plane crashed in Dakar and all on board were killed.

Examination of the plates from Bolivia quickly yielded ten more two-meson decays in which the secondary particle came to rest in the emulsion. The constant range of around 600 microns of the secondary meson in all cases led Lattes, Occhialini and Powell, in their October 1947 paper in *Nature*, to postulate a two body-decay of the primary meson, which they called π or pion, to a secondary meson, μ or muon, and one neutron particle. Subsequent mass measurements on twenty events gave the pion and muon masses as 260±30 and 205±20 times that of the electron respectively, while the lifetime of the pion was estimated to be some 10^{-8} s. Present-day values are 273.31 and 206.76 electron masses respectively and 2.6×10^{-8} s.

The number of mesons coming to rest in the emulsion and causing a disintegration was found to be approximately equal to the number of pions decaying to muons. It was, therefore, postulated that the latter represented the decay of positively-charged pions and the former the nuclear capture of negatively-charged pions. Clearly the pions were the particles postulated by Yukawa. This led to the conclusion that most of the mesons observed at sea level are penetrating muons arising from the decay in flight of pion created in nuclear disintegrations higher up in the atmosphere.

Powell was awarded the 1950 Nobel Physics Prize for his development of the emulsion technique and for the discovery of the pion; Occhialini was awarded the 1979 Wolf Prize (shared with George Uhlenbeck) for his contribution both to the pion discovery and to that of pair production with Blackett, who obtained the 1948 Nobel Physics Prize.

9

The Discovery of the Pion in Bristol in 1947

D. Perkins[1]

The year of the discovery of the pion was 1947, an interesting year for several reasons. That discovery was not the only new thing that happened in Bristol physics in 1947; and of course on a countrywide basis, the discovery of V-particles in Manchester also in 1947, was to my mind just as important. For these and other reasons I shall not restrict myself to the pion discovery, but tell you also of other events taking place at about that time, which form an essential background to the discovery of the pion.

First I want to point out how 1947 turned out to be an important year for particle physics in many ways. It was a sort of watershed. It occurred exactly halfway through the history of particle physics, which began of course exactly 100 years ago with the discovery of the electron by J.J. Thomson in 1897. In the following 50 years – to 1947 – some important progress was made – the discoveries of the positron, proton and neutron, neutrino, pion and muon and V-particles. Had we known it then, the evidence of the heavier flavours of both leptons and quarks had already been detected, (but of course not recognized). In those first 50 years, progress was slow, the particle physics community was very small, detectors were rudimentary, resources were meagre. Practically all published papers were signed by one, two or at most three authors (I recall my shock at seeing the first Bristol paper on pi-mu decay with **four** authors!) Communications with physicists overseas was quite difficult. One had to rely on the post: telephoning was precarious (especially between England and Italy) and very expensive.

---

[1] *N.E.*: Conference as part of the commemorations of the 50th anniversary of the pion discovery. Published in *Ciência e Sociedade* CBPF-CS-032/97, October 1997. The author is one of the pioneers in the research work that led to pion discovery.

The year 1947 was also a watershed in the sense that the pion and muon and V-particle discoveries stimulated an explosion of accelerator building: the subject thereafter moved into top gear, and for the first time, detailed and controlled experiments at accelerators began to take over a field which had so far been dominated by cosmic ray experiments where events were rare and you had to take what Nature gave you. Finally, for me personally, 1947 was important as the year in which I published my first paper.

The accompanying Table 1 lists some of the papers in particle physics appearing in 1947 or late 1946. Let us recall that in 1935, Yukawa had proposed a heavy quantum to account for the short-range nature of nuclear forces, with a Compton wavelength equal to this range and a mass of order 1/7 of the proton mass. Two years later, Street and Stevenson, and Anderson and Neddermeyer detected in cloud chambers the tracks of particles of mass intermediate between electron and proton – and thus called mesotrons. Yukawa had postulated that his quantum would decay giving an electron, and hence account for nucleon beta decay, and decay of the mesotrons was also observed in cloud-chamber experiments. But the problem was that even after numerous crossings of metal plates in cloud chambers, no mesotron had been seen to interact. Indeed, during the early 40 s, Japanese physicists had set up a "meson club" to study these questions. Some invented "weak coupling" pseudoscalar meson theories to account for this behaviour, while Sakata and Inoue proposed a 2-meson theory, in which a strongly-interacting Yukawa-type particle decayed to a weakly-interacting daughter mesotron. The English version of the Sakata-Inoue paper did not appear until late 1946. Marshak and Bethe, unaware of it and equally of the Bristol discovery of pi-mu decay in May 1947, re-proposed the two-meson hypothesis at the Shelter Island meeting in June 1947 (published in September 1947).

A crucial experiment on mesotrons was finally undertaken in Rome by Conversi and Piccioni, starting in 1943. They used a Rossi-type array (see Fig. 9.1) consisting of two iron blocks magnetized in opposite directions, which had the property of focussing particles of one sign of charge and defocussing those of the opposite sign. A particle stopping in the absorber block would be signalled by a coincidence of the Geiger trays CA, CB and CC and anticoincidence with the bottom tray, A. Any decay of the stopped particle was indicated by a delayed coincidence with counters D. Positive mesotrons stopping in the iron absorber decayed, while negative ones did not. They presumably underwent rapid nuclear capture, as expected for Yukawa particles and predicted by Tomonaga and Araki. I should mention that these experiments were running at the time of the Italian armistice in 1943. (I was at school at that time: I had just built my first radio using a crystal detector, and the news from Italy was the first thing I heard when I tuned in). The armistice was followed by the German army occupying Rome, the university being closed, and Conversi and Piccioni having to move the equipment (and themselves) to a safer place. Then the US Air Force started to bomb the place and they had to move once more (getting, Conversi told me, as close to Vatican City as they could).

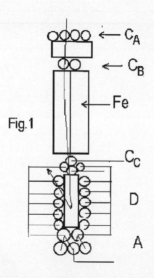

Figura 9.1: Rossi-type array used by Conversi, Pancini and Piccioni, 1943-47. The two parts of the iron block are magnetized in opposite directions, focussing particles of one sign and defocussing those of the opposite sign. A meson stopping in the absorber and decaying is given by the coincidence/anticoincidence of the various trays $C_A + C_B + C_C - A + D$ (delay).

After the war, the experiment continued, with Conversi and Piccioni joined by Pancini. They changed the absorber to carbon. The object of this was to record nuclear γ-rays which would follow nuclear capture of the mesotron and nuclear excitation and disruption; so they needed an absorber of a light element, which would not absorb the gammas. They changed to carbon: imagine their astonishment on finding that negatives stopping in carbon **all** decayed!

At the other end of Europe, in England, a by-product of the war and the nuclear programme was the setting up, in 1946, of a panel by the Ministry of Supply to oversee the development of special photographic emulsions to record nuclear particles. The chairman was Joseph Rotblat (winner of the 1995 Nobel Peace Prize) and the eight or nine of us on the panel included Cecil Powell, Otto Frisch, George Rochester and Berriman and Waller, the chemists from Kodak and Ilford. Under constant prodding and goading, by mid 1946 Ilford had produced a series of emulsions with four times the normal silver halide/gelatine ratio, which would record tracks of charged particles of ionisation down to about six times the minimum value. The series were called A, B, C$\cdots$ in order of increasing "grain" (= microcrystal) size, and B1, $\cdots$ in order of increasing sensitivity.

I was in the fortunate position at Imperial College, where I was a graduate student, that my supervisor, Sir George Thomson, was a Nobel Prizewinner and had been chairman of the famous Maud Committee in 1940, which had pronounced that a ^{235}U fission weapon would be possible. So he had a lot of clout, got on to the Air Ministry and persuaded them to arrange that flights of the RAF Photographic Reconnaissance Unit at Benson, near Oxford, should carry some of these emulsions for me (a total of six 3" × 1" 50 micron thick B1 emulsions). In November 1946 I got these back, processed them and found about 20 nuclear disintegrations, one of which was produced by an incoming charged particle (see Fig. 9.2). From scattering and ionization variation I estimated the mass to be 100-300 m_e. The secondary protons from the interactions were of low energy (4 or 5 MeV) which meant that (taking account of Coulomb barrier effects) this had to be capture in a light nucleus (C, N or O) of the gelatine, not Ag or Br or I of the halide.

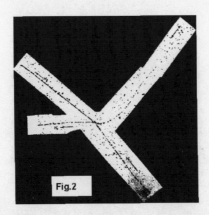

Figura 9.2: First negative meson capture event leading to disintegration of a light nucleus in the emulsion (B1 emulsion flown in aircraft from RAF Benson). Perkins: *Nature* **159**, 126 (1947).

I had heard about (but not seen) the Conversi result, that negative mesotrons stopping in carbon at sea-level in Rome seemed always to decay; while my negative particle, stopping in a light nucleus at 35 000' underwent nuclear capture. I realised there was a big difference here, but I had absolutely no idea what it all meant. Two weeks later, Occhialini and Powell in Bristol published six similar events.

The big breakthrough, however, was the publication, in May 1947, of two events in C2 emulsion exposed at the Pic du Midi, now called pi-mu decays, by Lattes, Occhialini, Muirhead and Powell – see Figs. 9.3 and 9.4. In Fig. 9.3, a parent particle comes to rest and decays into a second particle which leaves the emulsion surface just before coming to rest. The true secondary range could be quite well estimated. In Fig. 9.4, the event is complete; the secondary comes to rest after a range of 600μm. The estimated range in the first event and the observed range in the second event were almost exactly the same – evidence then for a simple 2-body decay. The two-meson hypothesis had finally been discovered by experiment.

In April 1997, at the conference dinner of an IOP meeting in Cambridge, someone asked me "what about the third pi-mu decay?" I had thought this was a

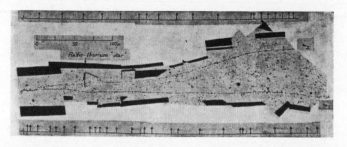

Figura 9.3: First $\pi \to \mu$ decay observed in C2 emulsion exposed at Pic du Midi. The secondary muon does not quite come to rest before leaving the emulsion surface (Lattes, Muirhead, Occhialini and Powell: *Nature* 159, 694 (1947)).

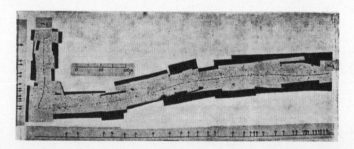

Figura 9.4: Second $\pi \to \mu$ decay observed by Bristol group. The muon comes to rest after a range of 610 μm.

50 year old secret, but I must have talked about it in an unguarded moment. Fig. 9.5 shows the event: it is a terrible picture of a complete pi-mu decay in B2 emulsion, exposed for me near Chamonix by Leprince-Ringuet. I found it in July 1947, and put it in my thesis. I did not publish it. In retrospect, it would of course have been independent confirmation of the Bristol events, from a different Laboratory and with different emulsions. But in those days, the atmosphere was very different from today: we didn't just rush into publication and I remember clearly having been deeply impressed by the Bristol work, and thinking that confirmation was not really necessary! I believe I did telephone Powell about it, but that was all.

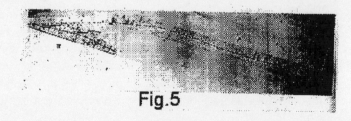

Figura 9.5: Second $\pi \to \mu$ decay observed by Bristol group. The muon comes to rest after a range of 610 μm.

In any case, the final proof of pi-mu two-body decay had to await until September 1947, after several dozen C2 plates had been exposed by Lattes from Bristol on Mt. Chacaltaya in Bolivia. Bristol found 10 more complete decays. Fig. 9.6 shows a histogram of the muon range distribution which, taking account of range straggling, clearly proves the two-body nature of the decay.

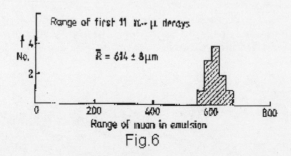

Figura 9.6: Histogram of ranges of muons in 11 complete $\pi \to \mu$ decays, proving that the decay is a two-body process (Bristol events of October 1947).

The Bristol people were able to show that, after taking account of geometrical efficiencies (the fact that the emulsions were only 50 μm thick and the muon range was 600μm), the true number of π^+ mesons and the number of negatives giving nuclear capture 'stars' were very comparable: thus the latter particles could be ascribed to π^-. Events were also found where π^+ and π^-, produced in nuclear disintegrations, came to rest in the same emulsion layer and underwent decay and nuclear capture respectively.

In 1948/49, both Kodak and Ilford were able to produce much more sensitive emulsions, NT4 and G5 respectively, which were sensitive to minimum ionizing particles. Fig. 9.7 shows four examples of complete $\pi \to \mu \to e$ decays. Since 1949, there have been no further developments of emulsion technology, and today's emulsions (made in Japan) are similar to those of 48 years ago.

The assumption that the first π–μ events were really decays was not taken lightly. A Bristol solid-state physicist, Charles Frank, looked into the question of whether such events could represent capture of a **negative** meson by an atom or molecule, which then catalyzed a nuclear reaction with release of energy and ejection of the original meson with a few MeV kinetic energy. Frank concluded that this could not occur in the emulsion, but **was** possible in a hydrogen-deuterium mixture.

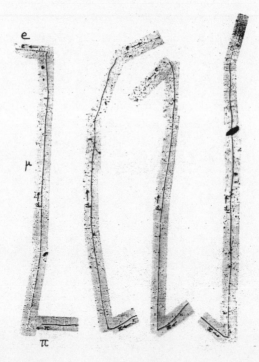

Figura 9.7: Four complete $\pi \to \mu \to e$ decays in G5 emulsion, showing the constancy of the muon range.

We now know this process as muon-induced fusion. A negative muon comes to rest in the hydrogen forming a μH^2 molecule (replacing one of the electrons because the muon binding energy is 200 times larger) and eventually finds an HD molecule to which it transfers (again, a reduced mass effect). Because the Bohr radius of the muon is only 10^{-11} cm, proton and deuteron can come close enough to fuse: $p + d \to {}^3\text{He} + 5.5$ MeV. The muon is ejected and can repeat the process, which was re-discovered experimentally a decade later by Alvarez at Berkeley (see Fig. 9.8). With the right HD mixtures, temperature and pressure, one muon can catalyze some 200 fusions. Unfortunately, this is just not enough — to create a negative muon. The problem is that the "sticking probability" of the muon to ${}^3\text{He}$ (about 0.5%) is just a little too big. But for this one wrong constant, muon-induced light element fusion could have been a viable source of power, and one would have had no Chernobyls, no problems with radioactive waste from fission reactors — and the standing of particle physicists on the world stage would have been a lot higher.

The final paper I want to mention from 1947 is that describing the neutral and charged V-particle events, by Rochester and Butler at Manchester — see Fig. 9.9. The first event (which incidentally, occurred on my 21st birthday) was probably what is now called $K_s^0 \to \pi^+ + \pi^-$, the second probably $K_{\mu 2}^+ \to \mu^+ + \mu_\nu$. After these two events, no other example was reported for two years; then confirmation

Figura 9.8: Hydrogen bubble chamber picture of HD $\rightarrow {}^3$He reaction catalyzed by negative muon capture into μHD molecule. The incident muon comes to rest, drifts as a neutral mesic atom, and is ejected with 5.4 MeV energy in the exoergic fusion reaction.

trickled in from MIT, CalTech, École Polytechnique. Personally, I believe that Rochester and Butler never received the acclaim due to them for this discovery. I recall that in 1952, a conference was held in Copenhagen to discuss particle and nuclear physics and, in particular, international collaboration in the field. At this time, the choice of Laboratory for CERN had not been made. Niels Bohr of course wanted it to be in Copenhagen, while Auger and Rabi preferred Geneva. (For obvious reasons, had we known then what we know today about Swiss banks during and after World War II, the decision may well have gone in favour of Copenhagen). George Rochester and I did the overnight sea crossing from Newcastle to Esbjerg. George and I went into the ship's bar and after several drinks, I persuaded him to sign a piece of paper to the effect that if he and Butler got the Nobel Prize, they would give me 10%. Unfortunately, like most of my schemes for making money, this has come to nothing. In any case, I've lost the piece of paper.

I should make some concluding remarks to indicate the atmosphere surrounding research in high energy physics 50 years ago. People did some very way-out experiments, simply for the hell of it, and because nobody really knew where or how the next big step would come. The early measurement of the charged pion lifetime provides an example of the ingenious approaches used. Table 2 shows some results. First, there was an experiment by Reg Richardson at LBL, measuring the number of pions surviving after successive spirals in a magnetic field. The pions were made at the 184" cyclotron, using an alpha-particle beam on a target (protons were no

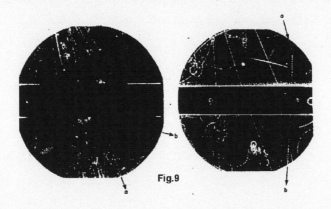

Figura 9.9: The year 1947 was rounded off by the publication of a neutral and a charged 'V-event' in a cloud chamber at Manchester. The upper neutral V on the right photo probably corresponds to $K_S^0 \to \pi^+\pi^-$, and the left one charged V, to $K^+ \to \mu^+ \nu$.

good, as they did not have enough energy: but the Fermi motion inside an alpha-particle gave the extra boost to get above threshold).

A second approach was that of Ken Greisen at Cornell: he looked for a "knee" in the cosmic ray muon spectrum. The idea was that, for pions of energy below 117 GeV (at which energy the pion decay length is equal to the scale height (6.5 km) of the atmosphere) decay is more probable than interaction. For higher energy, the reverse is true, and the pion decay probability varies as $1/E$. Hence the (negative) index of the muon energy spectrum from pion decay becomes one greater than that of the pions, so this "knee" measures the pion lifetime.

The third approach was that of Camerini and others at the Jungfraujoch. I remember Powell describing this in an evening lecture at Bristol. He pointed out that the pion lifetime was a few nanoseconds, and therefore delicate and sensitive apparatus would be necessary. So saying, he reached under the lecture bench and produced a cocoa tin! The method employed was to stick a vertical pole into the Aletsch glacier, and tie to it at different heights, a number of cocoa tins containing nuclear emulsions. Measurement of the relative numbers of upward-moving pions and muons surviving to different heights then gave a measure of the lifetime.

Finally, Martinelli and Panofsky repeated the Richardson experiment. It will be seen that all four experiments got the wrong answer, by between 5 and 9 times the stated errors; but their average is not so very far from the value accepted today!

Since those days of half a century ago, experimental particle physics has undergone profound changes. The detectors employed are incomparably more complex and sophisticated, the teams necessary to operate them run to 100 s of people instead of 3 or 4, and, worst of all, a battle for funds is being continually fought with one's fellow scientists. Everyone is expected to give "value for money" in a field where the eventual values of basic research cannot possibly be predicted.

But some things have not changed at all. Fifty years ago, Cecil Powell described his feelings on finding all those wonderful new processes in nuclear emulsions. He said it was "as if, suddenly, we had broken into a walled orchard, where protected trees flourished, and all kinds of exotic fruits had ripened in great profusion". Well, the walled orchards still exist today. Perhaps they are not so easy to find, but they **are** there, and it is for the new generation of physicists to find them, as I am sure they will.

Nov. '46	Sakata Inoue	*Prog. Theor. Phys.* **1**, 143	2 meson hypothesis
Jan. '47	Perkins	*Nature* **159**, 126	First 'σ-star' (π^-)
Feb. '47	Conversi, Pancini, Piccioni	*Phys. Rev.* **71**, 209	Negative mesotrons decay in carbon (μ^-)
Feb. '47	Occhialini Powell	*Nature* **159**, 186	6 'σ-stars'
May '47	Lattes Occhialini Muirhead Powell	*Nature* **159**, 694	2 π-μ decays
Sept. '47	Marshak Bethe	*Phys. Rev.* **72**, 506	2 meson hypothesis (again)
Oct. '47	Lattes, Occhialini Powell	*Nature* **160**, 453	644 mesons 105 σ-stars 11 complete π-μ 499 ρ-mesons (μ)
Oct. '47	Frank	*Nature* **160**, 525	Meson-induced fusion (μ HD)
Oct. '47	Rochester	*Nature* **160**, 855	*V*-particles

Table 1. Papers on Meson Physics 1946-7.

	Method	Result (nanosecs)
Richardson (1948)	Decrease in intensity of pions spiralling in field (Berkeley SC)	8 ± 2
Greisen (1948)	Knee in cosmic ray muon spectrum	60
Camerini *et al.* (1948)	Intensity of upward travelling muons above glacier	6 ± 3
Martinelli & Panofsky (1950)	Richardson method	19.7 ± 1.4
Present Value		$26.03 \pm .02$

Table 2. Early Measurements of Charged Pion Lifetime.

Bibliography

[1] S. Abachi *et al.* – DØ Collaboration, *Phys. Rev. Lett.* **74**, 2422 (1995).

[2] Proceedings of the "Hadron Colliders at the Highest Energy and Luminosity" – Ed. by A.G. Ruggiero – World Scientific (1996) and references therein.

[3] Collections of notes and talks of the Very Large Hadron Collider Physics and Detector Whorshop. March 13-15 (1997).

[4] The practice of the internationalism in High Energy Physics is an old idea. Everything is showed by several cultures separating several countries around one well defined purpose: "to probe the matter in all its aspects".

[5] C.D. Anderson, *Phys. Rev.* **43**, 491 (1933); S.H. Neddermeyer and C.D. Anderson, *Phys. Rev.* **51**, 884 (1937).

[6] C.M.G. Lattes, G.P.S. Occhialini and C.F. Powell, *Nature* **160**, 453 (1947) and Part II, *ibid.* p. 486; C.M.G. Lattes, H. Muirhead, G.P.S. Occhialini and C.F. Powell, *Nature* **159**, 694 (1947).

[7] K. Nin *et al. Prog. Theo. Phys.* **47**, 280 (1972).

[8] S. Abachi *et al.* – DØ Collaboration – *Phys. Rev. Lett.* 74, 2422 (1995).

10

Discovery of Pi and Mu Mesons and Brazil-Japan Collaboration on Cosmic-Rays

Yoichi Fujimoto

It is my great honor to be invited at the Annual Meeting of Sociedade Brasileira de Pesquisadores Nikkeis, and to be given an opportunity to talk on the research of cosmic-rays. As was mentioned by the speech of the President, Professor Yashiro Yamamoto, this year is the fifty years since the discovery of pi and mu mesons by Professor Cesar Lattes. His discovery is very important to Japanese physics, giving the decisive evidence for Yukawa meson theory and Sakata two meson model. So, there started close friendship between physics communities of the two countries, particularly for those on particle physics and cosmic-rays. From this basis, we are working in collaboration on cosmic-ray study with experiment at the mountain laboratory at Bolivian Andes, the historic place of the pi and mu meson discovery. Professor Lattes is the comandante of our cosmic-ray collaboration running since 1962.

Discovery of pi and mu mesons

Let me first show the historic events of pi and mu meson, which were observed by the photo-graphic plates of nuclear emulsion exposed at Pic du Midi in France (Fig. 10.1). The transparency shows mosaic of the microscopic picture of the meson tracks registered in the photographic plates. Here, you can recognize that a slow cosmic ray meson comes from outside and stops in the middle of photographic emulsion layer, and, from the stopping point, a new meson of low energy appears. In the historic paper of Lattes, Muirhead, Occhialini and Powell published in *Nature*, 1947, they call it the double meson event, and they reported the two examples of this type.

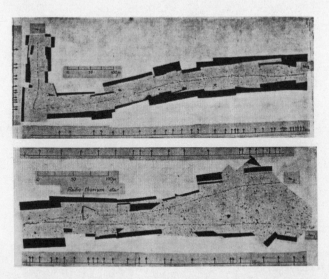

Figura 10.1: The first double-meson events found in nuclear emulsion plates. Lattes, Muirhead, Occhialini and Powell; *Nature*, 159 (1947), 694. It is a mozaic of the microscopic picture of particle track registered in nuclear emulsion layer of a photographic plate. Left is the first, and is the second event they found.

They thought of a possibility that a negative meson comes to stop in emulsion and is absorbed by a nucleus, and, after some nuclear transformation, a meson receives some amount of energy and goes out of the nucleus. At the same time, they did not forget a possibility of a heavy meson coming to rest in emulsion and decays into a light meson under the two meson hypothesis. Of course, it was not possible to draw a definite conclusion from observation of two examples.

They knew immediately that what they observed in their photographic plates would have decisive significance on the particle physics. Lattes convinced Powell and Occhialini, to make an immediate exposure of the photographic plates to cosmic-rays at higher altitude. He knew that the exposure is possible at Chacaltaya meteorological observatory in Bolivia, at altitude of 5 200 m, near city of La Paz. He went alone there with the photographic plates, and made the exposure with help of Ismael Escobar, who was running mini-observatory there. After the exposure of a month, Lattes made test processing at house of Escobar in La Paz, and he saw a good example of the double meson event under his microscope. Now the expedition was successful.

After his coming back to Bristol with exposed photographic plates, the processing was made and the scanning of meson events was carried out by a whole of Bristol group. In this way, they obtained dozens of the double meson events. What is remarkable is that range of the second meson, from its start to the end, is always constant, about 600 microns in the photographic emulsion, showing the second meson is emitted always with the same amount of kinetic energy. It shows clearly the two-body decay of a heavier meson (pi-meson or pion) into a lighter meson (mu-meson or muon) and an undetected neutral particle (neutrino).

At the same time, they found events of a stopped meson of other types. In the figure, you see an example of stopping meson making nuclear disintegration. It is a negative pi-meson stopped and absorbed by a nucleus giving its rest energy into the nucleus and making its disintegration.

Nuclear emulsion technique

It is known that the photographic plates are a good detector of radiation, as you remind the old discovery of Uranium-radioactivity by Becquerel by use of the photographic plates. Tracks of individual alpha-particles from radio-activity were first observed by Kinoshita, a Japanese studying in Rutherford Laboratory, in 1910s. A large improvement of the photographic emulsion, suitable for observing tracks of a particle, was made by the effort of Waller in Illford Co., just after the World War II. He succeeded to increase the content of silver bromide crystal in emulsion nearly by one order of magnitude. Here, you can see the difference in comparison of tracks of alpha-particles from radioactivity. New photographic plates with improved emulsion record a particle with much thicker track (Fig. 10.2). The photographic emulsion, particularly made for the purpose, is called as the nuclear emulsion, and the plate with such emulsion is named as the nuclear emulsion plate.

The nuclear emulsion plates could register a slow proton with a clear track, and it was used as a proton detector for study of nuclear reactions. It was the original intention for the improved emulsion.

Young school of São Paulo University at Bristol

Occhialini was the first to join Powell in Bristol, when Powell was setting up a laboratory for study of nuclear reactions with the improving nuclear emulsion technics. Then, Occhialini immediately realized that this new emulsion technics opened a large possibility for cosmic-ray studies as well. He suggested Powell to invite Cesar Lattes, and later Ugo Camerini, to Bristol from São Paulo as a key

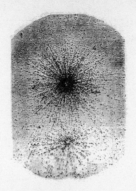 

Figura 10.2: Microscopic picture of tracks of alpha-particles which come out from a radio-activity source placed at the center. Left: Tracks in an ordinary photographic plate by Kinoshita and Ikeuti (1915). Right: Tracks in a nuclear emulsion plate, Ilford B, by Occhialini and Powell (1947).

member of the new group with new emulsion technique.

You know Occhialini was together with Gleb Wataghin at Universidade de São Paulo, working on cosmic-ray showers with a cloud chamber. Wataghin laboratory of Universidade de São Paulo was the first nuclear physics laboratory in Brazil, and many leaders in theoretical and experimental modern physics in Brazil were old boys from this laboratory. Occhialini had to leave the laboratory during the wartime, because of his Italian passport. Lattes was working there on slow mesons with his cloud chamber. He sent to Occhialini a picture of cosmic-rays taken by the cloud chamber to show him the progress. Then, Occhialini replied to Lattes with a microscopic picture of tracks of particles in the new nuclear emulsion plate at Bristol. Lattes noticed the excellent quality of tracks registered in the nuclear emulsion plates, and he decided immediately to join Bristol group and asked Occhialini to send the invitation.

I can find in Occhialini writing, how he wanted young and active Brazilians joining his venture into cosmic-ray study with the new technique, and how they worked in Bristol bringing their young energy and new culture into the traditional and academic atmosphere of English university.

So, Lattes came to Bristol in the beginning of 1946. First, he worked on nuclear reactions with the new plates, and he extended his study to neutron detection with Boron-loaded nuclear emulsion plates. The Boron-loaded plates of his idea worked very well for neutron detection, and, at the same time, he found it is stronger against the fading of latent images than the unloaded ordinary plates.

Immediately after, Lattes and Occhialini decided to make an exposure of the emulsion plates at a high mountain, that is Pic du Midi in France, trying to observe cosmic-ray particles. Occhilanini is a mountain-climber, and he supported his life as a mountain-guide at Itatiaia during wartime. So, it was not difficult for him to go there with nuclear emulsion plates of ordinary type and of Boron-loaded type.

After processing the exposed plates at Bristol laboratory, they found under the microscope, that the Boron-loaded emulsion plates registered many more tracks than the normal emulsion plates, showing that the Boron plates are really strong against fading. In a large number of events of cosmic-rays recorded in the Boron plates, they saw rich varieties of cosmic-ray phenomena, which were never seen before. Above all, there was this two-meson event. Now, the study of cosmic-ray mesons with Boron plates was put with highest priority.

Now you see that a young school of São Paulo, being planted in Bristol with the new nuclear emulsion technique, opened their flower of two meson discovery.

Powell, the head of laboratory, knew that this venturing into the field of cosmic-ray study was possible and successful only with participation of young school of São Paulo University. He then became enthusiastic for international collaboration and cultural exchange. His laboratory had been since then full of young physicists from all over the world. I worked there from 1953 to 55, and I benefited well the international atmosphere of Bristol laboratory.

Meson theory from Japan

Let me talk a significance of the two meson discovery, that is, how the long-lasting puzzles of meson theory were solved by the discovery and how it opened a new stage of particle physics.

Yukawa original intention of his theory of meson, published in 1935, is really ambitious. By introducing a new particle, with mass intermediate between an electron and a proton, he wanted to solve the two fundamental problems of nuclear physics. One is the nuclear forces, which bind protons and neutrons together into a nucleus. He assumed that the nuclear force was due to exchange of a virtual meson between a proton and a neutron, The other is beta-decay, a radioactive nucleus decaying with emission of an electron and a neutrino. He proposed that this is decay of a meson, virtually emitted from a nucleus, going into an electron and a neutrino. The newly proposed particle was called as U-particle in his first paper, but later the name was settled down to meson showing its intermediate mass.

It is an important question to me why this idea came out with a young unknown physicist in Japan, but not with the experienced authorities at the center of physics, that is, in Europe. Please remind you that the situation of Japan in 1930s was miserable and nothing good for the basic research. Though there were already some pioneer works on modern physics, Japanese academic community was weak and sick. The universities were under threat of the thought-police and the dominating militarism.

Naturally, his new theory was met by cool reaction from national as well as international communities. It was fortunate for him, because he had good collaborators around him. Among them, I might mention Shoichi Sakata and Mituo Taketani. Through their collaboration, they were able to formulate his meson idea into a form of quantum field theory of mesons, and start to make the theoretical calculations and the quantitative comparison with experimental information. There was already the quantum theory of electrons and of electro-magnetic field, but it was not known how to generalize the scheme to cover possible cases of a new particle with intermediate mass.

A part of answer to the previous question can be found in Taketani methodology, the three-stage theory, which he formulated from his study on history of sciences. The work was almost at the same time with Yukawa idea of meson. In European physics communities, it was generally believed that the science develops through two stages, the phenomenological stage and the essentialistic stage. It means that the former is with empirical formulas and the latter with the complete theory. Taketani pointed out that there is an important intermediate stage, called the substantialistic stage. It concerns to find out which is the key substance of the phenomena in question. He got his idea from the fact that the discovery of a neutron was the key to understand the phenomena of atomic nuclei. He was firmly convinced that the Yukawa idea is just this substantialistic theory that is needed to arrive at the essentialistic theory of sub-nuclear world.

Mesons in cosmic-rays

The cosmic-rays, which was discovered already in 1912, were beginning to be understood only in 1930s with new detector technics of counter-controlled cloud chamber under strong magnetic field. The positron was discovered by Anderson in 1932 giving an evidence for Dirac theory of an electron. Then, there was the discovery of electron-positron pair production by Blackett and Occhialini.

The cosmic-rays were known, phenomenologically, to be composed of the two components. The soft component, which is absorbed by lead plate of a few cm thick, was found mainly electrons, positrons and gamma-rays. While, there was the hard component that can easily penetrate through lead plates of 10 cm or more. In 1936, Anderson and Neddermeyer obtained an evidence for existence of a new particle with intermediate mass, *i.e.*, a meson. Since then, a large part of the hard component was known to be mesons.

The discovery of a meson in cosmic-rays was certainly encouraging to Yukawa and his collaborators in Japan. At the same time, it turned attention of physics communities in Europe and America to the study of the new particle, meson. Through intensive studies on the meson problem, both from the side of theory and of cosmic-ray experiments, the meson theory of Yukawa was found with full of difficulties. The meson problem was full of puzzles. It was the general situation around 1939-40, when the break of Second World War destroyed international exchange of scientific information.

From the theoretical study, the meson theory was found with inherent difficulties, though many of which were common to the case of electrons and of electro-magnetic field. The comparison with cosmic-ray experiment was more serious. Yukawa theory of mesons could not give even qualitative agreement with the experimental results on cosmic-ray mesons. Even Yukawa himself was becoming skeptical to his theory, thinking that the frame of field theory might be not suitable for this original idea and he looked for a new theory beyond the quantum theory. It was Shoichi Sakata, who is confident in the original meson idea, and tried to find a solution under the hypothesis of two mesons. He told me that he wanted to solve the problem with the substantialistic term of Taketani methodology. Sakata, Tanikawa and Inoue published, in 1942, their result of analysis with the two meson theory. This work was not informed to outside of Japan, and it was made public over all the world only in 1947. What Lattes found with the nuclear emulsion plates was just identical to the case studied by Sakata and Inoue.

New picture with two mesons

Let me summarize the understanding of mesons after discovery of two mesons, pi and mu. The pi-meson, or pion, is a particle which Yukawa introduced to understand the nuclear forces. Protons and neutrons bind together to make a nucleus, because they exchange pi-mesons, Therefore, a pi-meson has a strong interaction with a nucleus, and it is one of what we call hadrons. A pi-meson decays, with short life-time, into a mu-meson and an undetectable neutral particle (mu-neutrino). Therefore, we can find pi-mesons in cosmic-rays only just after its creation by nuclear collision of cosmic-rays. A positive pi-meson, when it comes to stop, will make decay into a positive mu-meson. A negative pi-meson, after its stopping, will be absorbed by a nucleus and make the nuclear disintegration (Fig. 10.3).

Mu-meson is cosmic-ray meson which Anderson and Neddermeyer found in their cloud chamber. Though a mu-meson has mass a little less than that of a pi-meson, they are very much different. Mu-meson has only weak interaction with a nucleus, several order of magnitude smaller than that of pi-mesons. So, a mu-meson is nothing to do with nuclear forces, and, in a way, it is similar to an electron. It is one of what we call leptons, and may be better to call a heavy electron rather than a meson. A mu-meson decays into an electron with continuous energy spectrum, so the decay is associated with emission of more than two undetectable

Figura 10.3: Microscopic picture of a meson event of another type. It is the first event, discovered by Perkins in 1947, which shows a negative meson coming to stop in the emulsion layer and making the nuclear desintegration with emission of three charged particles.

neutral particles (electron-neutrino and mu-neutrino). Since life-time of a mu-meson is much larger than that of a pi-meson, it occupies the major part of the hard component of cosmic-rays.

Discovery of artificial production of mesons

Lattes left Bristol at the end of 1947 and went to Radiation Laboratory at Berkeley, California, trying to find the artificial production of mesons. There was 184-inch cyclotron, the biggest in the world, which is producing a beam of alpha-particles with 380 MeV. Since an alpha-particle is made of two protons and two neutrons, the beam is equivalent to a beam of protons with one quarter of the energy, 95 MeV. This energy is not enough to create a pi-mesons in its collision with a proton at rest in the target. Lattes knew that protons and neutrons in a nucleus are moving just like molecules in gas. He can estimate the motion of protons and neutrons in a nucleus and he knows the best estimated mass value of a pi-meson. He concludes that the artificial production of a pi-meson is possible with this alpha-particle beam, if you take into account of the motion of a proton and a neutron inside a nucleus.

He collaborated with Gardner there to find positive and negative pi-mesons artificially produced by irradiation of alpha-particle beam on the target. His knowledge on the nuclear emulsion technics and his estimated value is essential in the design of experiment. They arranged the produced pi-mesons to be deflected by the magnetic field and to stop in the nuclear emulsion plates after passing through the absorbing material layer. The pi-mesons were found exactly at the place he anticipated.

In the next year, 1949, just before leaving Berkeley, he found artificial pi-mesons produced by high energy gamma-ray from 300 MeV electron synchrotron there.

This work at Berkeley was enough to open doors of accelerators to meson physics. It is the beginning of particle physics with large accelerators, which we call high energy physics. The physics communities became enthusiastic for construction of large accelerators for study of pi-mesons, and then for studies on variety of new particles.

If he had not gone to Berkeley, of course, the accelerator experiment of mesons would be certainly delayed, and I do not know who would be the pioneer of high energy physics neither how would be the fate of large accelerators. Instead, he would benefit further improved nuclear emulsion plates with high sensitivity to detect an electron track. There is a microscopic picture of the improved nuclear emulsion plates which record chain decay of a pi-meson to a mu-meson, and then to an electron (Fig. 10.4).

Figura 10.4: The pi-mu meson decay events in the improved nuclear emulsion plate, Ilford G5. The nuclear emulsion is sensitive to all charged particles including electrons. One can recognize that a pi-meson decays into a mu-meson and then decays into an electron.

Japan, Brazil and Nikkei colonies

Yukawa received Nobel Prize in 1949, after his idea was fully confirmed by the two meson discovery. It was a big national event for Japan after the defeat.

There was movement among Nikkei colonies to invite Yukawa to São Paulo at his occasion. They collected the donation from Japanese immigrants for the fund to invite Yukawa here, but unfortunately visit of Yukawa was not realized because of his health condition. Instead, they decided to use the fund to contribute to the theoretical physics group in Japan, encouraging their research activity, and the money was sent through Mainichi Press. In the old document, we find that the donated sum was as large as a few million yen at that time, very large amount for the physics research under miserable economical situation in Japan. I can find name of Shigueo Watanabe as a member of the committee of this movement.

At the receiving side in Japan, Professors Tomonaga and Taketani were responsible since Yukawa was in USA. The donated money was used for scientific publication of internal circular on the particle physics and for discussion meeting for preparing Japanese work to be presented at the international conference on theoretical physics held in 1953 at Kyoto. This Kyoto conference was the first international conference on science after the defeat. Those were not all. Theoretical physics professors thought that they should encourage experimental work on meson physics, more specifically, the study on cosmic rays with the nuclear emulsion technics. They were impressed by much by the two meson discovery, and they wanted, someday, nuclear emulsion work in Japan with fruitful results. They gave a part of Brazilian Nikkei fund to young experimentalists on cosmic rays. I am one of those who received the support, and I was able to buy binocular attachment to my monocular microscope. With this microscope, I saw the two meson decay event in the nuclear emulsion plates made by Fuji Film and Sakura Film.

Taketani came to São Paulo by invitation of Instituto de Fisica Teórica at Pamplona. This year, 1958, happened to be the anniversary of 50 years of Japanese immigration. Yukawa visited São Paulo at this occasion, and both professors expressed their deep gratitude to Nikkei members for their fund to help the particle physics research sent a several years ago. They went together to a small village called Mizuho, where the movement started to help Japanese physics.

A close friendship between Japanese and Brazilian physicists started at this occasion. Since then, there is a continuous exchange of researchers between the two countries, and the number increases year after year.

Collaboration experiment on cosmic-rays at Mt. Chacaltaya

Our cosmic ray collaboration work is based on such history. After coming back home from study at Bristol with the nuclear emulsion technics, I attended a workshop held at Yukawa Institute (Institute of Fundamental Physics) in Kyoto in 1956 trying to draw a future plan of cosmic-ray experiment on Inter-University basis. We came to a new idea of detector, which is a multi-layered sandwich of lead plates and nuclear emulsion plates to catch high energy electron showers, and we named the new device as the emulsion chamber. With this idea of emulsion chamber, a group of young boys planned a balloon experiment for observation of production of pi-mesons in high energy cosmic ray nuclear collisions. The experiment was made immediately with good success.

Seeing the emulsion chamber works well, we planned to increase the size and to work at higher energy region. The balloon was not suitable for the purpose, and we started the mountain experiment at Norikura observatory at altitude of 2 800 m. We found the idea working well, but the mountain laboratory was to low to get enough number of observed events. So, we were looking for possibility to make the experiment at famous Chacaltaya observatory, the place of two meson discovery. Yukawa was kind to us to introduce our intention to Lattes, asking the experiment at Chacaltaya to be the collaboration of Japanese group and Brazilian group. I remember well that we were very glad to receive a positive answer.

Lattes visited Japan in 1961 to attend the international cosmic ray conference at Kyoto, and at that occasion we got together to make the plan for collaboration experiment. In those days, there were no way to make international collaboration experiment in our Japanese system, so we have to ask most of the necessary expenditure abroad to be on Lattes side. He was very generous and he promised us to prepare every necessary material at Chacaltaya, and offered traveling and staying expenses. He told us to bring Japanese emulsion plates and X-ray films prepared for Norikura exposure, then the experiment could start immediately. Following his advice, Yokoi and myself went, in 1962, to his laboratory at São Paulo University together with photographic material, and we started the emulsion chamber experiment at Chacaltaya.

Since then, we obtained a number of experimental results. What we found on meson production in high energy cosmic ray nuclear collisions was later confirmed by the large accelerator experiment at CERN.

Besides, we encountered with a new phenomenon that we named Centauro. It is an arrival bundle of high energy hadrons, which can not be expected from the present knowledge from the accelerator experiments. It could be that, among cosmic ray particles arriving at earth from space, there are particles not known yet in the laboratory experiment. We will wait another occasion in future to talk on more details of the cosmic ray collaboration experiment.

11

César Lattes: um dos descobridores do méson π

José Maria Filardo Bassalo[1]

Comemoram-se, neste ano de 1989, os 65 anos de idade do físico brasileiro Cesare Mansueto Giulio Lattes, nascido em Curitiba, capital do Estado do Paraná, em 11 de julho de 1924, cujo nome encontra-se citado em qualquer livro sobre Partículas Elementares escrito (e que se venha a escrever) no mundo, por haver ele descoberto (juntamente com Muirhead, Occhialini e Powell) uma das partículas fundamentais da natureza: o píon.

Muito embora haja nascido em Curitiba, Lattes iniciou o então Curso Primário, em 1929, no Instituto Menegati, em Porto Alegre, havendo, contudo, completado esse Curso na Escola Americana, em Curitiba, de 1931 a 1933, após estudar o ano de 1930, em uma Escola Pública de Torino, na Itália. Já o então Ginásio foi cursado por Lattes no Instituto Médio "Dante Alighieri", em São Paulo, no período 1934 a 1938. Por fim, o seu Curso Superior ele o realizou na Universidade de São Paulo, havendo recebido o Grau de Bacharel em Física, em 1943, pela então Faculdade de Filosofia, Ciências e Letras (FFCL) dessa Universidade.

A motivação de Lattes para o estudo da Física, deveu-se a dois fatos [1]. Primeiro, ao terminar o Ginásio, soube que o Professor tinha três meses de férias por ano, ao invés de um mês, como ocorria com a maioria das demais profissões, o que o deixou bastante interessado. Segundo, como a maioria das matérias que estudou no Ginásio eram do tipo "decoreba", à exceção de Física e Matemática das quais gostava e aprendeu, Lattes inclinou-se para o estudo da Física devido ao incentivo

---

[1] *N.E.*: Publicado originalmente na série *Ciência e Sociedade* CBPF-CS-001/90.

que lhe deu seu pai Giuseppe (que, inclusive, falou sobre seu filho com Gleb Wataghin, um físico ítalo-russo, que veio a São Paulo para montar o Departamento de Física da FFCL) e seu professor Luís Borello, que ensinava a disciplina "Ciências Físicas e Matemáticas" no "Dante Alighieri".

Na FFCL da USP, Lattes foi aluno de Marcello Damy de Souza Santos, em Física Geral e Experimental: de Abrahão de Morais, em Física-Matemática: de Giácomo Albanese, em Geometria Projetiva; e de Wataghin e Giuseppe P.S. Occhialini, em disciplinas profissionais do Curso de Física. Enquanto as aulas desses três primeiros professores eram mais tradicionais, isto é, no sentido de serem nelas estudados assuntos de Física já consagrados nos livros-textos, as de Wataghin e Occhialini eram, principalmente, baseadas em seminários sobre temas publicados em Revistas Especializadas em Física, graças à excelente biblioteca que Wataghin organizou e manteve sempre atualizada, na FFCL.

Muito embora Lattes haja sido aluno de Occhialini apenas no 3° ano de seu Curso, em 1943, e somente em uma disciplina sobre raios-X, a maneira curiosa como ministrou essa disciplina, permitiu que Lattes apreendesse bastante sobre a leitura de filmes de raios-X uma vez que, sendo Lattes o único aluno de Occhialini, as aulas deste naquela disciplina consistiam em fazer seu aluno destrinchar os filmes de raios-X que revelava. (Aliás, existe um fato curioso sobre Occhialini. Ele havia vindo para o Brasil em 1938, chamado por Wataghin, porque era (e ainda é) um excelente físico experimental, pois fizera com Patrick Maynard Stuart Blackett a célebre experiência em 1932/1933 que comprovou a existência do pósitron descoberto por Carl David Anderson, em 1932 e, além do mais, era anti-fascista numa Itália dominada pelo fascismo. No entanto, quando o Brasil entrou na II Guerra Mundial (1942), Occhialini foi considerado inimigo de nosso país por ser cidadão italiano. Em virtude disso, foi ser guia de montanhas do Parque Nacional de Itatiaia, chegando, inclusive, a escrever um Guia para Turistas. Esteve também no Rio de Janeiro, contratado por Carlos Chagas para o Instituto de Biofísica. Contudo, em 1943, voltou a trabalhar na USP).

Logo depois de formado, Lattes começou a trabalhar em pesquisas com Wataghin, Mário Schenberg e Walter Schützer, resultando desse trabalho três artigos publicados nos *Anais da Academia Brasileira de Ciências* (Volume **4** (XVII), p. 269 (1945), com Wataghin), na *Physical Review* (Volume **69**, p. 237 (1946), também com Wataghin) e ainda nos *Anais da Academia Brasileira de Ciências* (Volume **3** (XIX), p. 193 (1947), com Schenberg e Schützer).

Quando Occhialini voltou para a Inglaterra em 1944, deixou para Lattes uma câmara de Wilson que não funcionava. Lattes, então, consertou-a (com a ajuda de Andrea Wataghin e Ugo Camerini), fez chapas fotográficas e mandou para Occhialini que trabalhava com Cecil Frank Powell, no H.H. Wills Physical Laboratory, da Universidade de Bristol. Sendo Assistente da Cadeira de Física Teórica e Física Matemática da FFCL (primeiro como 3° assistente (1944-1945) e depois como 2°, a partir de 1945), entusiasmou-se por haver colocado em funcionamento a câmara de Wilson. Decidiu-se, então por Física Experimental (já que havia trabalhado também em Física Teórica), e pediu que Occhialini o levasse para trabalhar em Bristol, para onde partiu no começo de 1946, a convite de Powell, incorporando-se dessa forma, ao grupo de pesquisa liderado por esse físico, grupo esse que veio a se tornar o famoso Grupo de Bristol, responsável pela descoberta dos píons, conforme veremos a seguir.

Em Bristol [2,3], Powell e Occhialini começaram a trabalhar com espalhamento elástico entre prótons e nêutrons, numa região em torno de 10 MeV de energia, com um novo tipo de emulsão nuclear construída pela Ilford LTd. (Ilford C2 Nuclear Emulsion), muito melhor que as emulsões tradicionais, já que era muito mais concentrada. Aí, então, foi dada a Lattes a tarefa de trabalhar na calibração dessa nova emulsão, ou seja, converter o alcance de prótons e de partículas alfa, em energia, e determinar o fator de contração (shrinkage factor) da emulsão depois de revelada. Os prótons usados por Lattes foram obtidos no acelerador Cockcroft-Walton de Cambridge através das seguintes reações: $D(d,p)_1H^3$, $_3Li^6(d,p)_3Li^7$, $_3Li^7(d,p)_3Li^8$, $_4Be^9(d,p2n)_4Be^8$, $_5B^{10}(d,p)_5B^{11}$ e $_5B^{11}(d,p)_5B^{12}$. (A notação $X(x,y)Y$ indica a reação $x+x \rightarrow y+Y$). Esse trabalho foi feito por Lattes juntamente com o físico francês P. Cuer e com o estudante inglês P.H. Fowler (neto do célebre Ernst Rutherford, o descobridor do núcleo atômico) e resultou em dois artigos publicados, respectivamente, na *Nature* (**159**: 301(1947)) e nos *Proceedings of the Physical Society*, London (**59**: 883 (1947)). (Antes disso, Lattes havia feito outros trabalhos com Cuer, relacionados com a desintegração do samário (C.M.G. Lattes and P. Cuer, *Nature*, **158**: 1947 (1946) e C.G.M. Lattes, E.G. Samuel e P. Cuer. *Anais da Academia Brasileira de Ciências*, **1** (IXI): 1 (1947)).

Depois desses trabalhos, Lattes se interessou em calcular a energia dos nêutrons, independentemente de saber a direção dessa partícula ao ser espalhada pelo próton. Desse modo, Lattes carregou uma chapa de emulsão com compostos de boro (que a Ilford havia preparado para ele) e a colocou na direção de um feixe de nêutrons decorrentes da reação $_5B^{11}(d,n)_6C^{12}$. O boro, então, ao receber o nêutron transformou-se em duas partículas alfa e num hidrogênio três, segundo a reação: $_0n^1 + _5B^{10} \rightarrow 2\,_2He^4 + _1H^3$. Desse modo, Lattes conseguiu medir a energia e a direção do nêutron. Contudo, esses nêutrons obtidos em laboratório tinham energia da ordem de 13 MeV e Lattes queria nêutrons mais energéticos.

No outono de 1946, Occhialini entrou de férias e foi passear nos Pireneus. Lattes, então, solicitou-lhe que levasse chapas (cerca de duas dúzias, cada uma com 1cm × 2cm, com aproximadamente 50 mícrons de espessura de emulsão) com boro e sem boro para ficarem expostas aos raios cósmicos, no observatório francês localizado no Pic du Midi, de Bigorre, nos Pireneus, durante o período de seu lazer. Na mesma noite em que Occhialini regressou a Bristol, ele e Lattes revelaram as chapas. No dia seguinte, Powell juntou-se aos dois e começaram a examiná-las. Logo perceberam que as placas com boro apresentavam mais eventos que as sem boro, uma vez que este composto químico tinha o PH certo para manter a imagem latente por mais tempo. Desse modo, o grupo de Powell [4] (agora composto de H. Muirhead, R.M. Payne e U. Camerini) iniciou um árduo trabalho para estudar esses eventos. Assim, após alguns dias de busca ao microscópio, uma jovem moça microscopista, de nome Marietta Kurz, encontrou um raro evento interpretado pelo grupo como um duplo-méson, isto é, um traço grosso (correspondendo a um méson–stopping meson) e no seu fim emergia um segundo méson com cerca 600 mícrons de alcance todo contido na emulsão. Alguns dias depois, um segundo méson duplo foi encontrado, no entanto, o segundo méson da dupla não parou na emulsão; porém, estudando a ionização que provocou, foi possível extrapolar um alcance também de aproximadamente 600 mícrons. Em vista disso, esses primeiros resultados sobre os mésons-duplos foram publicados na *Nature* (**159**: 694 (1947)), num artigo assinado por Lattes, Muirhead, Occhialini e Powell. Aliás, no volume **159** dessa mesma revista (p. 331) há um outro trabalho assinado apenas por Lattes e Occhialini, no qual apresentam o cálculo da direção e energia dos nêutrons oriundos dos raios cósmicos. (É oportuno salientar que foi fácil ao Grupo de Bristol identificar essas partículas como mésons e não como prótons, isto devido à variação da densidade de grãos com o alcance e, também, devido ao espalhamento ser muito largo).

Como havia alguns grupos de pesquisas na Europa trabalhando com emulsões expostas em aviões [5] (inclusive D. Perkins, do Imperial College, Londres, um pouco antes, havia encontrado um evento semelhante ao do Grupo de Bristol, trabalhando com novas emulsões), e acreditando haver descoberto um processo fundamental (apesar de haver possibilidade desses resultados serem explicados por uma reação exotérmica do tipo $\mu +_a X^b \to _{a-2}X^b + \mu^+$), o Grupo de Bristol precisava urgentemente de mais eventos de mésons-duplos além dos 1,5 já obtidos. Assim, Lattes foi ao Departamento de Geografia da Universidade de Bristol e viu que havia uma estação meteorológica no Monte Chacaltaya (que no idioma aimara significava "adonde los huesos tiemblan") [6], de 5 500 m (mais alto que o Pic du Midi, que tem 2 800 m) e a uma distância de 20 km, por terra, de La Paz, capital da Bolívia. Então, de posse de várias chapas de emulsão carregadas com boro, Lattes deixou Londres num avião brasileiro e viajou para a América do Sul. (é oportuno salientar que a sorte (ou seu anjo-da-guarda) ajudou Lattes, pois o Diretor do H.H. Wills, Prof. Tyndall, queria que ele tomasse um avião inglês, que caiu em Dakar matando todos os ocupantes). Depois de expor por um mês essas placas, Lattes partiu de volta a Londres, com escala no Rio de Janeiro. Já na Bolívia e no Rio, um rápido exame dessas placas, evidenciou um terceiro méson-duplo e que tinha o mesmo alcance do primeiro evento obtido, ou seja, aproximadamente 600 mícrons. (É oportuno esclarecer que, no Rio, Lattes usou o microscópio de Joaquim Costa Ribeiro, professor da Faculdade Nacional de Filosofia (FNFi), para confirmar esse evento. Ao mostrá-lo a José Leite Lopes e Guido Beck, também professores da FNFi, estes acreditaram, como Lattes, tratar-se da produção de mais um méson-duplo. Leite Lopes passou, então, a estudá-lo teoricamente). Chegando em Bristol, Lattes mostrou esse terceiro evento a Blackett, e o Grupo começou imediatamente a procurar novos eventos, encontrando por fim, cerca de mais 30 [7].

Convencidos de que haviam descoberto um processo fundamental da Natureza, Lattes, Occhialini e Powell passaram a calcular as massas desses mésons primário e secundário contando os grãos deixados nos traços. Desse modo, encontram uma massa de 139 MeV/c^2 para o méson primário, e 106 MeV/c^2, para o méson secundário. O méson-duplo foi então identificado como sendo devido a um processo de decaimento de méson de Yukawa (o méson primário ou pi) no mésotron de Anderson (o méson secundário ou mi) e em mais uma partícula neutra de massa aproximadamente nula, presumivelmente um neutrino, necessário para o balanço do momento. Esses cálculos foram publicados na *Nature* (**160**: 453 a 459 e 486 a 492 (1947) e nos *Proceedings of the Physical Society, London* (**61**: 163 (1948)).

Antes de prosseguirmos com a descrição do trabalho científico de Lattes, é necessário situarmos a Física das Partículas Elementares [8,9] nas décadas de 1930 e 1940, para compreendermos a importância dessa descoberta. Em 1934, Enrico Fermi propôs a existência de uma nova força na Natureza – a força fraca – para poder explicar o decaimento beta (n $\rightarrow$ p + e + ν). Logo depois, em 1935, Hideki Yukawa propôs um novo tipo de partícula – méson – para poder explicar a razão pela qual prótons e nêutrons (nucleons) se confinam no núcleo, sem que os prótons sofram repulsão coulombiana. Para Yukawa, portanto, uma nova força na Natureza – a força forte – é que mantém os nucleons confinados no núcleo, força essa de curto alcance (10^{-15} m) e resultante da emissão e absorção de "elétrons pesados" (os mésons, com massa da ordem de 150 MeV/c^2), por parte dos nucleons. Por outro lado, em 1936, Anderson e Seth Henry Neddermeyer, ao fazerem fotografias de raios cósmicos em câmara de Wilson, descobriram novas partículas de massa intermediária entre a do elétron e a do próton, denominadas por eles de mésotrons. Em consequência, a questão que então surgiu foi a de saber se os mésotrons andersonianos eram os mésons yukawianos [10].

As experiências com os mésotrons andersonianos, contudo, continuaram na década de 1940. Logo em 1941, Louis Leprince-Ringuet e colaboradores (S. Gorodetstky, E. Nageotte e R. Richard-Foy) determinaram a massa dessas partículas como sendo da ordem de 120 MeV/c^2. Ainda nesse mesmo ano de 1941, Franco Rasetti determinou sua vida média como sendo de 2×10^{-6} s. Todavia, a presença de mais de um méson na radiação cósmica havia sido demonstrada na célebre experiência realizada no Brasil, em 1939, por Wataghin, Marcello Damy e Paulus Aulus Pompéia, e conhecida mundialmente como *chuveiros penetrantes* (*cascade showers*). Em vista disso, uma série de físicos teóricos (inclusive os brasileiros Schenberg e José Leite Lopes) começaram a formular teorias sobre os dois mésons [11,12]. Dentre elas, destacam-se os trabalhos realizados, entre 1935 e 1947, por vários grupos de pesquisas de físicos japoneses liderados pelo próprio Yukawa, por Shoichi Sakata, por Sin-Itiro Tomonaga e por Mituo Taketani, nos quais a ideia básica era a de que os mésotrons seriam produto do decaimento dos mésons de Yukawa. (Note-se que essa mesma ideia foi apresentada por H.A. Bethe e R. Marshak, em 1947). Neste mesmo ano de 1947, um resultado experimental importante foi obtido pelos físicos italianos Marcello Conversi, Ettore Pancini e Oreste Piccioni, qual seja, o de que os mésotrons andersonianos interagiam fracamente com os núcleos, o que não deveria acontecer com os previstos por Yukawa. (Os primeiros resultados obtidos por esses físicos italianos foram durante a II Guerra, e quando Roma foi bombardeada eles tiveram que continuar suas pesquisas clandestinamente em um porão do Liceo Virgilio, próximo do Vaticano. Registre-se ainda, que esse resultado foi confirmado por T. Sigurgeirsson e A. Yamakawa, em Princeton, também em 1947).

Voltemos a Lattes. Conhecedor do trabalho de Conversi, Pancini e Piccioni [13] (mas não dos japoneses, pois a Guerra dificultou a comunicação científica) e de posse do resultado da experiência de seu Grupo, em Bristol, Lattes pensou numa maneira de produzir artificialmente os mésons-pesados pi. Por essa época, isto é, em 1947, começou a operar o sincro-ciclotron de 184 polegadas, em Berkeley, California, que acelerava partículas alfa a 380 MeV. Lattes imaginou então produzir esses mésons pesados nesse acelerador. Pois bem, então, por indicação de Wataghin, ele foi contemplado com uma bolsa de estudos da fundação Rockfeller. Porém, esse acelerador fora construído com verbas da Comissão de Energia Atômica dos Estados Unidos da América e, portanto, o acesso a ele não era muito simples, tendo em vista o pós-guerra. Então, em uma de suas viagens ao Rio de Janeiro, Lattes e Leite Lopes foram falar com o Almirante Álvaro Alberto da Mota e Silva, representante do Brasil na Comissão de Energia Atômica das Nações Unidas, para ver se ele conseguia permissão da Comissão de Energia Atômica dos Estados Unidos, para Lattes trabalhar naquele acelerador, a qual foi conseguida por intermédio de Mr. B. Baruch. Contudo, antes de seguir para a Califórnia, o descobridor do píon esteve em Copenhague, por convite do Nobel Niels Bohr, e também na Suécia, para falar sobre seu trabalho em Bristol.

Muito embora soubesse que partículas alfa de 380 MeV (95 MeV/nucleon) eram insuficientes para produzir os mésons pi, Lattes pensava contar com a energia de Fermi (movimento interno) dos prótons e nêutrons componentes desse tipo de partículas, para a produção desejada. (Esse cálculo foi checado por Lattes e Leite Lopes). Com efeito, em fins de 1947 e com essa ideia em mente, partiu para Berkeley e, em 1948, juntamente com Eugene Gardner, produziu artificialmente os primeiros mésons pi negativos (*Science*, **107**: 270 (1048); *Proceedings of the American Physical Society* (1948)). Mésons pi positivos também foram produzidos por Lattes, Gardner e John Burfening (*Physical Review* **75** (3): 382 (1949). (Burfening era aluno de Doutoramento de Lattes, assim como o foram S. White, K. Bowker, S. Jones e F. Adelman). Nessas experiências, os mésons π tiveram suas massas estimadas em cerca de 300 massas do elétron ($m_e = 0,5$ MeV/c^2), resultado esse publicado em um artigo assinado por Walter H. Barkas, Gardner e Lattes, (*Proceedings of the American Physical Society* (1948)). Essa descoberta teve grande repercussão nos Estados Unidos e também no Brasil, graças à divulgação na imprensa feita por Leite Lopes.

Depois de uma breve vinda ao Brasil para ser o Patrono dos Químicos Industriais da Escola Nacional de Química no final de 1948, Lattes continuou em Berkeley. Assim, no começo de 1949, com o síncrotron de elétrons de 300 MeV construído por Edwin Mattison McMillan, Lattes detectou cerca de uma dúzia de mésons $\pi^{\pm}$ (bem como a primeira evidência do píon neutro (π^0)), produzidos por fótons. Esse trabalho, no entanto, ficou inédito. É oportuno observar que Ernest Orlando Lawrence, o inventor do ciclotron, de início, negou-se a acreditar na produção artificial dos mésons pi, porém, logo depois convenceu-se. A existência dos mésons π, deu a Yukawa o Prêmio Nobel de Física de 1949, e a Powell, o de 1950. Estranhamente não foram laureados Sakata, Occhialini e Lattes. No entanto, eles pertencem a uma galeria – os *Nobéis Injustiçados* –, da qual fazem parte nomes famosos na Física como Ludwig Edward Boltzmann, Lord Kelvin, Paul Langevin, Arnold Sommerfeld, Ernst Pascual Jordan, dentre muitos outros.

Após esse sucesso em Bristol e em Berkeley, Lattes volta ao Brasil para materializar a ideia da criação de um Centro de Pesquisas Físicas, no Rio de Janeiro, já que, ainda em Berkeley, havia conversado com Nelson Lins de Barros, irmão do Ministro João Alberto (que era um político altamente influente no Brasil), sobre a viabilidade dessa sua ideia. Assim, em dezembro de 1948, no Rio, Lattes, juntamente com Leite Lopes, foi visitar o Ministro João Alberto para falar sobre aquela sua ideia. Este, auxiliado pelos irmãos Nelson e Henry, tornaram então possível, legal, material e financeiramente, o Centro Brasileiro de Pesquisas Físicas (CBPF). Para poder estruturar este Centro, Lattes contou com o prestígio e a colaboração dos matemáticos Antonio Aniceto Monteiro, Leopoldo Nachbin e Francisco Mendes de Oliveira Castro e dos físicos Elisa Frota Pessôa, Gabriel Fialho, Jayme Tiomno, Lauro Xavier Nepomuceno, além, é claro, do próprio Leite Lopes. Desse modo, instalado no Edifício Rex, localizado na Rua Álvaro Alvim, Cinelândia e no 21° andar, e ainda sob a Secretaria Geral de Nelson Lins de Barros, começou o CBPF, hoje um orgulho brasileiro e respeitado internacionalmente. (Não pode deixar de ser registrado o fato de que quando o Ministro João Alberto não pôde dar mais o seu próprio dinheiro (por motivo de doença) para o CBPF, este sobreviveu graças ao auxílio dado por Euvaldo Lodi, então presidente da Confederação Brasileira da Indústria, por Mário Almeida, dono do Banco Comercial, (este doou 500 contos de réis com os quais se construiu o primeiro prédio do CBPF, no campus da Universidade do Brasil, na Av. Wenceslau Braz, 71), e por uma subvenção mensal dada pelo presidente Getúlio Vargas).

Conforme vimos, Lattes deixou o Brasil e viajou para a Inglaterra vinculado à FFCL da USP e ligado à regência da Cadeira de Física-Teórica e Física-Matemática. Pois bem, na volta ao Brasil, no começo de 1949, Wataghin tentou mantê-lo em São Paulo, ao criar uma Cadeira para ele. Mas como era uma Cadeira sem nome, sem verbas e sem salas, Lattes pediu demissão e foi ao Rio para assumir o cargo de Professor Titular do CBPF e, também, a Cadeira de Física Nuclear da Faculdade Nacional de Filosofia da Universidade do Brasil que Joaquim Costa Ribeiro e Leite Lopes haviam criado para ele [14,15].

Do final da década de 1940 até a metade da década de 1950, Lattes ocupou-se com a criação e a consolidação de grupo de pesquisa em Física, quer em São Paulo, quer no Rio. Assim, foi Diretor Científico do CBPF, trabalhou na elaboração da criação do Conselho Nacional de Pesquisas e, a partir de sua instalação em 1951 (graças ao grande empenho do Almirante Álvaro Alberto), foi Membro de seu Conselho Deliberativo e, por fim, juntamente com o Professor Ismael Escobar, implantou o Laboratório de Física Cósmica da Universidad Mayor de San Andrés, na Bolívia. (Neste último trabalho, Lattes contou com a colaboração decisiva de Occhialini, Marcel Schein, Alfred Hendel e R. Escobar).

Em virtude de um grande escândalo de corrupção envolvendo o CNPq e o CBPF, Lattes desgostou-se e aceitou o convite para ser responsável (*Head*) pelo grupo de Emulsões Nucleares do "Institute for Nuclear Studies Enrico Fermi" da Universidade de Chicago, para onde partiu em junho de 1955, ficando ali até novembro de 1956. Dessa data em diante, trabalhou como Professor Associado do College of Science, Literature and Arts da Universidade de Minnesota, USA, até dezembro de 1957. Nessa duas Universidades, Lattes fez estudos sobre a desintegração eletrônica dos mésons pi e a correlação angular do decaimento $(\mu - e)$ de mésons produzidos por raios cósmicos em grandes altitudes (30 000 m), estudos esses que resultaram em vários trabalhos (H.L. Anderson and C.M.G. Lattes, *Il Nuovo Cimento* **VI** (6): 1356 (1957); P.H. Fowler, P.S. Freier, C.M.G. Lattes, E.P. Ney and S.J. St. Laurant, *Il Nuovo Cimento* **VI** (1): 63 (1957), **III** (28) (1957), C.M.G. Lattes and P.S. Freir, *Proceedings of the Padova Conference on Elementary Particles*, **5**: 17 (1957); P.H. Fowler, P.S. Freier, E.P. Ney, P.H. Perkins and C.M.G. Lattes, *Il Nuovo Cimento* **4** (9): 1 (1959)).

Depois desse período de trabalho nos Estados Unidos, Lattes voltou ao Brasil e, por insistência de Schenberg e de José Goldemberg, passou a trabalhar na USP, em tempo parcial, exercendo a Cadeira de Física Superior do Departamento de Física da FFCL, havendo organizado o seu Laboratório de Emulsão Fotográfica. Nessa época, começo de 1960, Lattes recebeu uma proposta do professor Schein, de Chicago, onde trabalhara, para tomar parte em uma organização internacional para estudar cerca de 5% dos 100 litros de emulsão que haviam sido expostos em balões, a 30 000 m de altitude. Desse modo, surgiu o *International Cooperation Emulsion Flight* – ICEF, do qual faziam parte grupos de pesquisa dos Estados Unidos, Canadá, Dinamarca, Inglaterra, França, Alemanha, Itália, Polônia, Suíça, Japão, Índia e o Brasil, naturalmente. Desse ICEF resultaram alguns importantes trabalhos, sendo que um deles foi apresentado na Conferência Internacional sobre Raios Cósmicos, ocorrida em Kyoto, Japão, em 1961. Também como resultado das pesquisas realizadas pela ICEF, Lattes preparou sua Tese para o Concurso de Professor Catedrático de Física Nuclear da FNFi (que havia sido criado por ato do presidente Eurico Gaspar Dutra), posteriormente editada pelo Núcleo de Estudos e Pesquisas Científicas (NEPEC), do Rio de Janeiro, em 1962. (Aliás, até o presente momento, a UFRJ ainda não marcou a data desse Concurso).

Como titular da Cadeira da Física Superior da FFCL da USP, Lattes organizou um grupo de pesquisas para estudar os fenômenos produzidos pela interação de raios cósmicos de energia superior a 10^{12} eV [16], em câmaras de emulsão-chumbo expostas no Monte Chacaltaya. Encerrada, em 1962, a participação do Brasil no ICEF, Lattes começou nesse mesmo ano a organizar o Projeto de Colaboração Brasil-Japão [16,17], para estudar também essas emulsões. Aliás tal Colaboração deveu-se a uma feliz coincidência, já que os físicos japoneses Y. Fujimoto, S. Hasegawa, M. Koshiba, J. Nishimura, K. Niu, M. Oda e K. Suga, do Institute for Nuclear Studies, em Tokyo, haviam organizado um projeto para examinar a produção de múons em Chacaltaya, apoiado por Bruno Rossi, do Massachusetts Institute of Technology (MIT), projeto esse denominado *Bolivian Air Shower Joint Experiment* – BASJE. A Colaboração Brasil-Japão ocorreu graças aos contatos entre Yukawa (que visitou o Brasil em 1958), Mituo Taketani (que trabalhou no Brasil em 1958-1959, no Instituto de Física Teórica (SP) e em 1961-1962 na USP) e Lattes. Em 1967, Lattes transferiu-se para a UNICAMP a fim de organizar e dirigir o hoje Departamento de Geocronologia, Raios Cósmicos, Altas Energias e Léptons do IFGW. Assim, o Projeto de Colaboração Brasil-Japão passou a envolver o IFGW e o CBPF. Hoje, há a participação da Universidade Federal Fluminense, do Rio de Janeiro e, recentemente, da Universidade Federal do Mato Grosso. (Convém observar que Lattes, ainda no Departamento de Física da FFCL da USP, em 1962, começou a fazer os primeiros estudos sobre Cosmologia e Cronologia Teórica, havendo mesmo, em 1964, em Pisa, na Itália, implantado laboratórios de datação por fissão espontânea do urânio).

Esse Grupo de Colaboração Brasil-Japão teve a participação de importantes físicos japoneses (além dos já citados), bem como ajudou na formação de vários físicos brasileiros, hoje bem conceituados no cenário científico internacional. Além de contribuir para a formação de físicos conforme destacamos acima, esse Grupo obteve resultados importantes como a descoberta de uma série de eventos novos, do tipo *bola-de-fogo* (*fireball*), eventos esses resultantes da interação de raios cósmicos com núcleos da atmosfera e observados na câmara de emulsão de Chacaltaya. Esses eventos são de dois tipos: formação do evento com produção múltipla de píons e formação do evento com produção de hádrons "exóticos" não-piônicos. Neste último tipo, temos eventos com pequena componente transversal do momento linear (p_t) (da ordem de 1 GeV/c) e com grande componente desse mesmo momento (da ordem de 10 GeV/c). Todos esses eventos foram comunicados em conferências e simpósios internacionais sobre raios cósmicos realizados nos quatro cantos do mundo, conforme veremos a seguir.

A ideia de *bola-de-fogo* foi apresentada por Wataghin [18], em 1941, por ocasião do Simpósio Internacional sobre Raios Cósmicos, realizado no Rio de Janeiro. Essa ideia tornou-se popular depois da interpretação estatística dada por Fermi, em 1950, e da interpretação hidrodinâmica dada por Lev Davidovich Landau, 1963. Por outro lado, Shunichi Hasegawa, em 1961, introduziu a natureza quântica da *bola-de-fogo* propondo o *H-quantum* (*Heavy-quantum: quantum pesado*) como a sua unidade básica. Contudo, o primeiro evento do tipo *bola-de-fogo* e com produção múltipla de píons observado pela Colaboração Brasil-Japão foi anunciado por Lattes na Conferência Internacional sobre Raios Cósmicos realizada em Jaipur, Índia, em 1963 (da qual foi Presidente da Sessão de Altas Energias), com o nome *Mirim* que na língua tupi-guarani significa *pequeno*, já que sua massa é entre 2-3 GeV/c². Em 1967, na Conferência Internacional sobre Raios Cósmicos ocorrida em Calgary, no Canadá, Lattes anunciou a detecção da *Açu* (*grande*, em tupi-guarani), um novo evento tipo *bola-de-fogo* com produção múltipla de píons e com massa entre 15-30 GeV/c². Já na Conferência Internacional sobre Raios Cósmicos levada a cabo em Hobart, na Tasmania, em 1971, um outro tipo de produção múltipla de píons foi anunciado por Lattes: *Guaçu* (*muito grande*, em tupi-guarani), com massa entre 100-300 GeV/c².

As produções de hádrons "exóticos" não-piônicos, a partir de um produto intermediário (*bola-de-fogo*) em uma reação nuclear, foram observadas pela colaboração Brasil-Japão na década de 1970.

Assim, esse Grupo comunicou na Conferência Internacional sobre Raios Cósmicos realizada em Denver, Colorado, em 1973 – o Centauro –, um evento tipo *bola-de-fogo* com a massa da *Guaçu*, porém com a componente transversal do momento linear (p_t) da ordem de 1 GeV/c, contra 400-500 MeV/c apresentada pela *Guaçu*. O nome *Centauro* foi dado por causa de um acontecimento estranho em sua produção. Nas outras famílias de *bola-de-fogo* (*Mirim*, *Açu*, *Guaçu*), a energia observada era maior na câmara de emulsão de cima do que na de baixo. No caso do *Centauro*, ocorre o contrário. É oportuno lembrar que, na Mitologia, O *Centauro* é um ser metade homem, metade cavalo). Por outro lado, na Conferência Internacional sobre Raios Cósmicos ocorrida em 1977, em Plovdiv, na Bulgária, Lattes anunciou um outro evento tipo *Centauro* que, contudo, por apresentar a massa da *Açu*, foi batizado com o nome de *Mini-Centauro*.

Por fim, a produção de hádrons "exóticos" não-piônicos [19], com p_t da ordem de 10 GeV/c – maior, portanto, do que o da família *Centauro* –, foi também observada pela Colaboração Brasil-Japão. O primeiro desses eventos foi apresentado por Lattes na *Topical Conference on Cosmic Rays and Particles Physics above 10 TeV* (promovida pela Universidade de Delaware e Bartol Research Foundation), em 1978, com o nome de *Geminion*, com massa da família *Mini-Centauro*, recebendo esse nome porque só produz um par de hádrons "exóticos". Um outro evento desse tipo, com a massa da família, porém como p_t da ordem de 10 GeV/c, foi comunicado pelo Grupo Brasil-Japão, no Wisconsin Symposium, em 1981. Esse evento recebeu o nome de *Chiron*.[2] Essa família *Chiron*, contudo, apresenta um aspecto novo em relação às famílias *Centauro* e *Mini-Centauro*, pois os hádrons secundários que ela produz apresentam uma produção múltipla de píons e um feixe de partículas de componente hadrônica e eletromagnética e batizada pelo Grupo de *Mini-Cluster*.

---

[2] Na mitologia, *Chiron* é um elemento da família *Centauro*.

Esses eventos tipo *bola-de-fogo* despertaram a atenção de outros pesquisadores no mundo, tanto que, no começo da década de 1970, começaram a busca desses eventos. Assim, grupos de físicos montaram câmaras de emulsão nas montanhas Pamir, na União Soviética, Fuji, no Japão, e Kambala, na China, constituindo, respectivamente, os *Grupos Pamir, Fuji e Kambala* [2]. O *Grupo Pamir*, por exemplo, composto de físicos soviéticos e poloneses, detectou, no início, apenas eventos do tipo *Mini-Centauro*. Porém, a partir de 1980, esse Grupo juntou-se ao Grupo Brasil-Japão e, em 1987 (*Physics Letters B*, **190**: 226), anunciou um provável candidato a *Centauro* observado nas montanhas Pamir. Os eventos tipo *bola-de-fogo*, com formação múltipla de píons, apresentaram confirmações em experiências realizadas em aceleradores (G. Pancheri and C. Rubbia, *Nuclear Physics A* **418**; 117c (1984); M. Jacob, *Nuclear Physics* **418**: 7c (1984); C. Rubbia, *Proceedings of the 1985 Internacional Symposium on Lepton and Photon Interaction at High Energy*, Kyoto, Japan: 242-273). No entanto, os eventos tipo *bola-de-fogo* com produção de hádrons "exóticos" ainda não foram evidenciados nos aceleradores, por ser a energia destes (da ordem 10^{11} eV) ainda baixa, comparada com a energia necessária para produzi-los.

Durante esse trabalho de pesquisa, Lattes preocupou-se com a formação de pessoas altamente qualificadas conforme já frisamos, preocupação essa comprovada pela série de Teses de Mestrado que orientou (E.M. Cioccheti, J.C. Hadler Neto, J.A. Chinellato, A.C. Fauth), de Doutorado (além das já citadas, acrescentam-se M.S. Mantovani, C. Santos, E.H. Shibuya, J.A. Chinellato, C.D. Chinellato, J.C. Hadler Neto), e di Laurea (G. Bigazzi, A. Michele), no Brasil, nos Estados Unidos e na Itália, quer na área de emulsões nucleares, quer na área de geocronologia.

Este perfil científico de Lattes mostra porque ele é o físico brasileiro mais conhecido no Brasil. Mostra, ainda, porque recebeu as maiores honrarias de todos os cantos do nosso país e, também, do mundo, durante esses 65 anos de vida. Dentre elas, destacam-se o título de *Doutor Honoris Causa* outorgado pela USP, em 1948, e recebido somente em 1964; o título de "Cavaliere di Gran Croce", *Ordo Capitularis Stellae Argentae Crucitae*, em 1948; o *Prêmio Einstein*, da Academia Brasileira de Ciências, em 1951; a Medalha de Ouro "Honra ao Mérito", da Rádio Nacional/ESSO, em 1951: o prêmio Ciências, do Instituto Brasileiro de Educação, Ciência e Cultura, em 1953; o *Prêmio Ernesto Fonseca Costa*, do Conselho Nacional de Pesquisas, em 1953, o título de *Cidadão Carioca*, em 1957; o título de *Cidadão Paulista Emérito*, em 1958 o título de *Personagem do Ano*, pelo Grêmio Cultural Rui Barbosa, em 1961; a *Ordem do Mérito Cultural*, da União Brasileira de Escritores, em 1969; o título de *Cidadão Honorário de La Paz*, Bolívia, em 1972; a *Medalha Carneiro Felipe*, do Conselho Nacional de Energia Nuclear, em 1973; o *Prêmio Moinho Santista* (SAMBRA) – Física, em 1975, a *Comenda Andrés Bello*, outorgado pelo Governador da Venezuela, em 1977; o *Prêmio Bernardo Houssay*, da Organização dos Estados Americanos, em 1978, os títulos de *Doutor Honoris Causa* e *Professor Emérito*, outorgado pela UNICAMP, em 1987, porém ainda não recebidos e, em 1987, o *Award in Physics of The Third World Academy of Sciences*, em Trieste, Itália.

Além dessas honrarias, Lattes foi escolhido patrono e paraninfo de várias turmas de Colandos das mais variadas profissões; é nome de logradouros (ruas e prédios) públicos no Paraná (Mandaquiri, Nova Esperança, Cambira), no Rio de Janeiro (Miguel Pereira, São Gonçalo, Nova Iguaçu). Por exemplo, o edifício onde se localiza o CBPF, tem o seu nome. Além do mais, é Membro Honorário-Titular-Fundador de várias Sociedades ou Academias Científicas do Brasil e do Mundo; proferiu Colóquios e participou de Seminários em várias Instituições Estrangeiras; foi Membro do Conselho da Sociedade Brasileira para o Progresso da Ciência, por vários mandatos: foi membro Consultor do Conselho Latino Americano de Raios Cósmicos; exerceu por três vezes o mandato de Membro da Comissão de Raios Cósmicos da União Internacional de Física Pura e Aplicada; foi, também, Membro Titular do Conselho da Sociedade Brasileira de Física; foi Membro da Comissão da Revista *Il Nuovo Cimento*, da Itália; exerceu vários cargos relacionados com ensino na UNICAMP, e possui verbetes na *Encyclopaedia Britannica*, na *Biographycal Encyclopaedia of Science and Technology*, de Isaac Asimov, na Enciclopédia Mirador Internacional, na Grande Enciclopédia Delta Larousse e na Enciclopédia Delta Universal.

Antes de concluirmos este trabalho sobre o professor Cesar Lattes, como é mundialmente conhecido, é oportuno fazer uma pequena referência à questão polêmica por ele levantada em 1979-1980, sobre a não constância da velocidade da luz. Nessa ocasião, Lattes fez experiência sobre a difração da luz e notou que a figura dela resultante se deslocava na medida em que o observador também se deslocava. Tal resultado, o levou a concluir que o mesmo violava o Princípio da Relatividade Restrita Einsteiniana. Essa conclusão foi por ele apresentada em uma reunião polêmica realizada na Academia Brasileira de Ciências, durante a qual se destacou Jayme Tiomno. Mais tarde, Lattes reconheceu que havia cometido um engano em sua interpretação sobre o resultado daquela experiência. Contudo, embora Lattes haja se enganado no varejo, talvez a sua intuição física indique que, no atacado, esteja certo. Segundo Mário Schenberg [21], uma compreensão física mais satisfatória da Teoria da Relatividade necessita de um maior aprofundamento do sentido físico dos conceitos topológicos nela envolvidos. Observa ainda Schenberg, que essa talvez seja a razão pela qual ainda não se conseguiu uma conciliação entre a Gravitação e a Teoria Quântica. Não teria, portanto, Lattes sentido que algo não vai bem com a Relatividade? Porém, ainda não sabemos o quê! Só o futuro apontará a direção certa pressentida por Lattes, acreditamos. (É importante chamar atenção para o fato de que o físico búlgaro Stefan Marinov no livro *Eppur si Muove*, escrito em 1977, relata algumas experiências por ele realizadas nas quais há evidências da violação do Princípio da Relatividade Restrita).

Em conclusão, gostaríamos de ressaltar dois aspectos da atitude científica do Professor Lattes. No começo da década de 1950, o então Ministro da Saúde estava interessado em acabar com a malária na baixada fluminense. Para isso, era necessário determinar o alcance médio do voo dos mosquitos transmissores. A técnica então utilizada era bastante rudimentar: pintar os mosquitos. No entanto, a pintura dificultava o voo e a análise do mesmo se tornava incompleta. Gustavo Oliveira Castro, biólogo-ecologista do Instituto Oswaldo Cruz falou a Lattes sobre esse problema. Lattes apresentou-lhe a solução que consistia na criação dos mosquitos numa solução bem diluída de acetato de tório, pois, ao nascer, o mosquito já estava com o tório e ficava, portanto, marcado pela radioatividade. Depois, concluiu Lattes, seria só acompanhar as partículas alfa emitidas pelo mosquito e determinar, desse modo, seu alcance médio. Foi um sucesso essa solução.

O segundo aspecto que ressaltaremos do espírito científico do Professor Lattes é a sua grande capacidade de organizar Grupos de Pesquisas. Além dos grupos por ele organizados e já citados aqui, em 1972, implantou o Grupo de Cronologia da UNICAMP e, em 1975, o Grupo de Detectores Eletrônicos para estudar os múons (e seus respectivos neutrinos) da radiação cósmica. Por fim, após se aposentar, em 1986, do cargo do Professor Titular da UNICAMP, reassumiu seu antigo posto no CBPF e na UFRJ. Está colaborando intensamente na formação de um grupo de Pesquisas do Departamento de Física da Universidade Federal do Mato-Grosso, (cujo convite para organizá-lo surgiu por ocasião do "III Encontro de Físicos do Centro-Oeste e Minas", ocorrido em 1985) e continua ativamente participando da Colaboração Chacaltaya-Pamir.

Agradecimentos

Agradeço aos professores Édson Hiroyuki Shibuya, do Departamento de Raios Cósmicos, Cronologia, Altas Energias e Léptons, do Instituto de Física "Gleb Wataghin", da Universidade Estadual de Campinas, e José Leite Lopes, Diretor do Centro Brasileiro de Pesquisas Físicas, pela leitura e crítica, a minha mulher, Professora Célia Bassalo, pela revisão e ao Serviço de Computação da Universidade Federal do Pará, pela edição em Microcomputador.

Bibiografia

[1] C.M.G. LATTES – Entrevista com Ricardo G.F. Pinto e Tjerk G. Franken (1976).

[2] C.M.G. LATTES – "My work in Physics with Nuclear Emulsions". In: *Topics on Cosmic Rays (60th Anniversary of C.M.G. Lattes)*: 1-5. Editora da UNICAMP (1984).

[3] C.M.G. – "My work in Mesons Physics With Nuclear Emulsion". In: *The Birth of Particles Physics*: 307-310. Cambridge University Press (1986).

[4] H. MUIRHEAD – "Encounters With Giulio Lattes". In: *Topics on Cosmic Rays (60th Anniversary of C.M.G. Lattes)*: 14-20. Editora da UNICAMP (1984).

[5] G.P.S. OCCHIALINI – "César Lattes: The Bristol Years". In: *Topics on Cosmic Rays (60th Anniversary of C.M.G. Lattes)*: 6-8. Editora da UNICAMP (1984).

[6] A. MARQUES – 25 anos da Descoberta do Méson pi. CBPF-CS-001/73 (1973).

[7] C.M.G. LATTES – "Leite Lopes and Physics in Brazil: A Personal Testimony". In: *Leite Lopes Festschrift*: 3-7. World Scientific. (1988).

[8] E. SEGRE – *Dos Raios-X aos Quarks*. Editora da Universidade de Brasília (1987).

[9] C.N. YANG – *Elementary Particles: A Short History of Some Discoveries in Atomic Physics*. Princeton University Press (1962).

[10] S. HAYAKAWA – "The Development of Meson Physics in Japan". In: *The Birth of Particles Physics*: 82-107. Cambridge University Press (1986).

[11] J.M.F. BASSALO – "Nota Histórica: A Contribuição dos Físicos Brasileiros para o Estudo dos Léptons". *Ciência e Cultura*, **38** (11): 1849-1858 (1986).

[12] J. LEITE LOPES – "Point-counterpoint in Physics: Theoretical Prediction and Experimental Discovery of Elementary Particles". In: *Topics on Cosmic Rays (60th Anniversary of C.M.G. Lattes)*: 27-47. Editora da UNICAMP (1984).

[13] O. PICCIONI – "The Observations of the Leptonic nature of the "Mesotron" by Conversi, Pancini and Piccioni". In: *The Birth of Particles Physics*: 222-241. Cambridge University Press (1986).

[14] J. LEITE LOPES – "Feynman e a Física no Brasil". *Ciência Hoje*, **9** (51): 72-73 (1989).

[15] A. SILVEIRA – A Física Moderna no Rio de Janeiro – Reminiscências. CBPF-CS-001/84 (1984).

[16] Y. FUJIMOTO – "Brasil-Japan Collaboration Experiment at Chacaltaya". In: *Topics on Cosmic Rays (60th Anniversary of C.M.G. Lattes)*: 45-61. Editora UNICAMP (1984).

[17] M. TAKETANI – "Professor Lattes e a Física no Japão", In: *Topics on Cosmic Rays (60th Anniversary of C.M.G. Lattes)*: 21-26. Editora da UNICAMP (1984).

[18] G. WATAGHIN – Entrevista com Cylon E.R. Gonçalves da Silva. *Ciência e Cultura*, **35** (11): 1712-1727 (1983).

[19] S. HASEGAWA – "Preliminary Results on Observation of Genetic Relation Among the Exotic Cosmic Ray Phenomena". In: *Topics on Cosmic Rays (60th Anniversary of C.M.G. Lattes)*: 71-80. Editora da UNICAMP (1984).

[20] S.A. SLAVATINSKI – "Investigation of the Fragmentation Processes of Projectible Hadrons and Jets Production with High P_t at Energy 10^4 TeV". In: *Topics on Cosmic Rays (60th Anniversary of C.M.G. Lattes)*: 99-121. Editora da UNICAMP (1984).

[21] M. SCHENBERG – "César Lattes: Grande Físico e Personalidade Extraordinária". In: *Topics on Cosmic Rays (60th Anniversary of C.M.G. Lattes)*: 11-13. Editora da UNICAMP (1984).

12

César Lattes

Edison Hiroyuki Shibuya

Cesare Mansueto Giulio Lattes ou Cesar Lattes, como é mais conhecido, é certamente uma das personalidades científicas brasileiras de reputação internacional. Suas contribuições foram decisivas para o nascimento da chamada 'Física das Partículas Elementares'.

No final dos anos 1940, com a observação do decaimento do píon em múon, Cesar Lattes estabeleceu de maneira inequívoca a existência do méson, tal qual preconizado por Hideki Yukawa. Essa partícula, responsável pela estabilidade do núcleo do átomo, foi denominada méson π (píon), que se transforma no múon e em outras partículas. Nessa ocasião, o múon já havia sido observado com algumas características que induziram a identificá-lo inicialmente como sendo a partícula de Yukawa, equívoco que foi devidamente demonstrado pelas célebres experiências de Conversi, Panchini e Piccioni. Após a observação de eventos de decaimento píon-múon, primeiro em Raios Cósmicos e posteriormente em máquinas aceleradoras de partículas (por exemplo o kaon), outras partículas foram sucessivamente sendo descobertas, principalmente em Raios Cósmicos, propiciando o surgimento da 'Física das Partículas Elementares'.

Nessa época, já se sabia que o átomo era divisível e constituído de partículas com carga elétrica positiva (prótons). Também haviam sido observadas, em 1932, por James Chadwick, partículas sem carga elétrica (nêutrons). Outras experiências, interpretadas por Ernest Rutherford, haviam indicado a existência de um núcleo atômico, onde os prótons estão concentrados e os elétrons espalhados em órbitas ao redor desse núcleo. Quando da descoberta do nêutron, o chamado modelo planetário de Rutherford já estava estabelecido e assim os prótons e nêutrons estariam compactados no núcleo. Como consequência deste modelo, surgiria uma nova questão: já que os prótons têm carga elétrica positiva, qual seria o agente responsável para anular as forças de repulsão entre as cargas positivas? Para responder a essa questão, Hideki Yukawa propôs a existência de uma partícula (méson) que seria o agente responsável pela coesão das partículas constituintes do núcleo. Portanto, a observação do decaimento do píon em múon foi de importância fundamental para o estabelecimento das forças nucleares e para o melhor entendimento do átomo. Em resumo, a observação desse fenômeno fez de Cesar Lattes um protagonista pioneiro no estudo dos constituintes nucleares do átomo.

No presente, as partículas outrora ditas elementares, com exceção do elétron, são encaradas como sendo constituídas de partículas 'mais elementares'. Assim, apesar
do píon (no momento conhecido como partícula, devido à 'quebra espontânea da simetria chiral do Bóson de Goldstone' – que expressão complicada!) atualmente ocupar uma posição secundária na classificação das partículas elementares, a sua observação e a sua descoberta envolveram aspectos de ideias lógicas presentes em todas as épocas, não só em Ciências, mas em todas as atividades humanas. São as ideias popularmente conhecidas como 'ovo de Colombo'.

Cesar Lattes conclui precocemente seus estudos universitários pois, por ocasião da conclusão do ginásio (atualmente 8ª série do $2°$ grau), a recém-criada Faculdade de Filosofia, Ciências e Letras – FFCL – da Universidade de São Paulo estava selecionando, através de exames vestibulares, alunos para o Departamento de Física, sob a responsabilidade de Gleb Wataghin. Lattes foi aluno também de Giuseppe Occhialini, que tendo retornado à Europa após o término da II Guerra Mundial e se ligado à Universidade de Bristol, convidou-o a participar de projeto para viabilizar o material fotossensível como um detector quantificador para traços de partículas atômicas e nucleares. Foi Lattes, dessa maneira, encarregado de determinar o fator de encolhimento (*shrinkage factor*) da nova placa de emulsão nuclear, muito mais concentrada que a antiga. Em consequência, essa tarefa propiciou determinar a relação alcance-energia dos prótons, relação de importância fundamental e que foi utilizada durante muitos anos. No mesmo experimento foram utilizadas placas impregnadas de Bórax (composto de Boro, na forma de $Na_2B_4O_7 \cdot 10H_2O$), segundo solicitação de Lattes à empresa fotográfica Ilford. Lattes planejava efetuar a reação: $n_0 + B_5^{10}O \rightarrow He_2^4 + He_2^4 + H_1^3$, objetivando determinar energia e momento dos nêutrons através destas placas. Vale observar que o nêutron é eletricamente neutro e, em assim sendo, não ioniza o material fotossensível, impossibilitando sua observação direta. A observação e a caracterização do nêutron (n_0) incidente no B_5^{10} se fazem produzindo 2 átomos de He_2^4 e 1 átomo de H_1^3.

O desaparecimento rápido da imagem fotográfica era extremamente crítico nos primórdios da técnica fotográfica. Os estudos de Lattes sobre esse efeito de perda da imagem latente foram de fundamental importância. As consequências dessa constatação podem ser avaliadas em um texto de Occhialini: Cesar Lattes, os anos de Bristol – Ciência e Cultura, **36** (11), novembro de 1948, p. 2067.

"Após a descoberta de que mesmo nas novas emulsões, o desaparecimento gradual da imagem era importante, ficou claro que o tempo de exposição das chapas que eu expus no Pic du Midi deveria ser encurtado. O ganho de algumas semanas causado por isso mostrou-se crucial. Naquela época, outros grupos da Europa expunham chapas em aeroplanos e nós, sem saber, participávamos da corrida do *decaimento píon-múon*".

Aproveitando uma viagem de férias de Occhialiani para os Pireneus (Pic du Midi e arredores), Lattes solicitou-lhe que levasse as placas de emulsão concentrada B1, algumas delas impregnadas com Bórax, para expô-las à Radiação Cósmica durante 1 mês. Os objetivos para com as placas normais, desprovidas de Bórax, estavam voltados para o estudo de raios cósmicos de baixa energia como também para verificar a possibilidade de detecção dos nêutrons da Radiação Cósmica.

Os resultados dessa exposição foram fundamentais para o desenvolvimento de métodos fotográficos para a Física, resultados resumidos por Lattes em 3 aspectos:

a) O Bórax, de alguma maneira, funcionou como estabilizador da imagem latente enquanto as placas desprovidas desse composto químico apresentaram significativo efeito de desaparecimento da imagem fotográfica;

b) As placas com Bórax apresentaram uma variedade de eventos com tal riqueza de detalhes que justificavam priorizar os estudos desses eventos. Apesar dessa priorização, os estudos sobre os nêutrons da Radiação Cósmica continuaram e seus resultados foram publicados nessa mesma ocasião;

c) Os eventos normais eram tão significativos que direcionaram as atividades do laboratório para as observações dos eventos de raios cósmicos de baixa energia.

Em suma, o Bórax se constituiu num grande achado para a consolidação das pesquisas em curso.

Desses trabalhos resultaram a observação, medição e caracterização do decaimento píon-múon. Entretanto, nessa exposição havia sido observado um evento completo e um outro apresentou o múon saindo da película fotossensível antes de atingir o repouso, ou seja, haviam sido obtidos tão somente 1,5 eventos, o que certamente é uma amostragem pobre. Tornava-se, portanto, urgente a rápida observação de mais eventos de decaimento píon-múon para se obter uma estatística razoável.

Uma montanha de fácil acesso viria a suprir essa necessidade, já que se sabia que o fluxo da Radiação Cósmica aumenta com a altitude. Com essas considerações, Lattes localizou o Monte Chacaltaya, que dista 20 km da cidade de La Paz, na

Bolívia. Exposições lá efetuadas mostraram 30 eventos de decaimento píon-múon. Medidos e analisados esses eventos, a massa do píon resultou em cerca de 300 vezes a massa do elétron, ou seja, em torno de 170 MeV/c^2, valor portanto maior que a energia de partículas α aceleradas pelo ciclotron de Berkeley, na Califórnia (energia de 380 MeV, ou de 95 MeV/nucleon).

Considerando que os nucleons (constituintes do núcleo atômico) não são estáticos, haveria possibilidade de colisão entre a partícula α incidente e um nucleon, este em movimento no sentido contrário à partícula incidente. Nessa condição favorável, dever-se-á somar a energia do nucleon àquela energia da partícula α. Feito o cálculo, Lattes verificou que a energia total era suficiente para a produção artificial do píon (energia total no sistema centro de massa das partículas em colisão um pouco maior que a massa de repouso do píon).

Com a estimativa da massa dos 30 píons observados, com a imagem do decaimento píon-múon observado em películas fotográficas expostas à Radiação Cósmica e com a expectativa da produção artificial do píon, Lattes embarcou para a Califórnia e, apenas uma semana após o desembarque, pode mostrar evidências da produção dessa partícula. Certamente a rápida observação da produção artificial do píon deve-se fundamentalmente ao conhecimento prévio que Lattes dispunha, ou seja, sabia o que procurar. Diz Lattes que quando ele chegou a Berkeley, havia píons registrados nas películas utilizadas previamente para a calibração do feixe do acelerador.

As medidas dos 30 píons decaindo em múons mostraram que estes percorrem uma distância média em torno de 600 micrometros dentro de película fotossensível antes de decaírem em outras partículas. Esse resultado e a produção artificial de píons poderiam ser utilizadas em aplicações práticas que requerem incidência de partículas que transmutam explosivamente em outras partículas. Estão em curso tentativas para fins médicos, irradiando células cancerosas localizadas, utilizando múons provenientes de feixe de píons, decaindo explosivamente sobre as células e consequentemente destruindo-as. O sucesso dessas tentativas certamente viria a ser motivo de satisfação para Lattes.

Anos mais tarde, Lattes inicia uma cooperação científica com pesquisadores japoneses, cooperação denominada Colaboração Brasil-Japão de Raios Cósmicos (CBJ). Essa cooperação continua até os dias atuais (1995) e, eventualmente, envolve também pesquisadores poloneses, russos, e de algumas outras repúblicas da ex-União das Repúblicas Soviéticas. Um dos resultados dessa cooperação foi confirmado recentemente pelo acelerador do CERN (atualmente denominado *Organisation Européenne pour la Recherche Nucléaire*) e esse resultado surgiu de exposições de grandes detectores providos de material fotossensível e placas de chumbo expostos à Radiação Cósmica incidente no Monte Chacaltaya. Essas exposições se constituem numa espécie de continuidade daquelas efetuadas quando da descoberta do decaimento píon-múon na Radiação Cósmica e a geometria utilizada para os detectores foi em consequência dos insistentes comentários de Lattes.

No interregno entre a descoberta da produção artificial (1947) e o início da Colaboração Brasil-Japão de Raios Cósmicos (1962), Lattes participou de outras experiências importantes, por exemplo, do I.C.E.F. (*International Cooperation on Emulsion Flight*). Apesar dessas experiências terem produzido resultados importantes, o autor dessas linhas imagina que os trabalhos desenvolvidos por Lattes na determinação das correlações angulares do decaimento píon-múon, obtidos numa série de experiências em Raios Cósmicos e também em aceleradores, ainda deverão ser melhor compreendidos, haja vista as recentes especulações teóricas sobre quebras de simetria chiral, levogira e destrogira.

O que se depreende da descrição dos decaimentos mencionados é o encadeamento dos resultados obtidos a partir da observação dos primeiros eventos de decaimento píon-múon em Raios Cósmicos. Com certeza absoluta a adição do Bórax às placas concentradas iniciais foi um fato que marcou todo o desenvolvimento subsequente. É claro também que não se sabia de antemão o efeito que essa adição produziria, estabilizando a imagem latente. Nesse sentido, pode-se dizer que a descoberta do píon na Radiação Cósmica foi por um acaso, porém um acaso que foi muito bem aproveitado, seja na utilização do Monte Chacaltaya, seja na busca da produção artificial dessa partícula. Assim fica evidente que a atuação de Lattes em todos esses episódios foi decisiva e que tudo é fruto do empenho e dedicação ao trabalho proposto. Verifica-se que todas as pessoas de sucesso atingem esse estágio, graças ao trabalho e determinação, e Lattes não é exceção. Parafraseando o próprio Lattes, pode-se dizer que em tudo que se faça é necessário algo como 10% de inspiração e 90% de transpiração para se fazer um trabalho bem feito.

Esse texto foi elaborado baseando-se em escritos do próprio Professor Cesare Mansuetto Giulio Lattes, seus amigos e colegas e nas suas descrições orais para diversas plateias, sendo, portanto, passíveis de erros de interpretação. Detalhes e trechos biográficos mais completos constam em publicações tais como:

1) "My work in meson physics with nuclear emulsions" de Cesar Lattes, republicado neste livro nas pp. 7-9.

2) *Topics on Cosmic Rays – 60th Anniversary of C.M.G. Lattes*, vol. 1,2, Editora da Universidade Estadual de Campinas, 1984.

3) *Cesar Lattes 70 anos – A nova Física Brasileira*, Alfredo Marques (Ed.), Rio de Janeiro, CBPF, 1994.

13

O Sucesso do Grupo de Bristol

Ana Maria Ribeiro de Andrade

O trabalho trata do processo que levou Cesar Lattes, Giuseppe Occhialini e Cecil Powell a descobrirem o méson-π na radiação cósmica, em 1947. Tendo como cenário o fim da Segunda Guerra Mundial, o estudo privilegia a participação do físico brasileiro Cesar Lattes. Visando integrar os elementos básicos da produção científica, esta construção histórico-social do méson-π examina as principais atividades de seus descobridores. Com o mesmo objetivo, destaca as atividades do H.H. Wills Physical Laboratory da Universidade de Bristol, os instrumentos de pesquisa utilizados por esses físicos, os aliados do grupo de Bristol no observatório do Pic du Midi e na estação meteorológica de Chacaltaya, o financiamento da pesquisa, a publicação dos resultados, a controvérsia científica, o sucesso e a premiação. O trabalho também mostra porque a Segunda Guerra Mundial propiciou o encontro desses três físicos e acentua que, ao lado do intenso trabalho em equipe no Wills Laboratory, da experiência anterior de Powell na utilização de chapas fotográficas na física nuclear e de Lattes e Occhialini em raios cósmicos, o sucesso foi alcançado devido ao aprimoramento da emulsão nuclear fabricada pela Ilford Ltd. Baseado em entrevistas, depoimentos, informações obtidas em arquivos privados e nos artigos publicados por esses físicos, o estudo exemplifica como o conhecimento histórico-sociológico pode contribuir para o entendimento das atividades daqueles que buscam desvendar os mistérios da natureza.

Lattes, Occhialini e Powell durante a Segunda Guerra Mundial

Tentando detectar partículas previstas teoricamente, Cesar Lattes [1] – professor assistente de Física Teórica e Matemática na Universidade de São Paulo – USP, desde 1944 – investigava um assunto de fronteira. Utilizava na pesquisa uma pequena câmara de Wilson conectada a dois contadores Geiger-Müller, dotados de dispositivos elétricos, e à qual estava acoplada uma câmara fotográfica. O equipamento havia sido construído em colaboração com Ugo Camerini e Andrea Wataghin, conforme o método desenvolvido pelo seu ex-professor Giuseppe Occhialini [2]. Lattes praticava uma física experimental de baixo custo – o estudo dos raios cósmicos – considerado então um setor da física nuclear de altíssimas energias. Essa era a linha de pesquisa da USP que interessava a físicos teóricos e experimentais. Além de compatível com o referencial teórico do grupo liderado por Gleb Wataghin [3] – a eletrodinâmica quântica – os trabalhos eram desenvolvidos em condições de igualdade com físicos de outros países. Os raios cósmicos estão na natureza, apresentando efeitos de altitude e intensidade um pouco diferentes, e, em decorrência disso, a pesquisa nessa área não requer grandes investimentos na montagem de laboratórios. Nos moldes dos institutos de física da Itália, de onde vieram Occhialini e Wataghin, os instrumentos eram construídos pelos próprios cientistas. Assim, o grupo da USP trabalhava nos mesmos problemas científicos e com métodos e meios semelhantes aos utilizados no exterior.

Experiências sobre a radiação cósmica foram realizadas em Campos do Jordão (SP) e nas minas de ouro de Morro Velho (MG), utilizando chapas de chumbo, contadores Geiger e circuitos eletrônicos com alto poder de resolução confeccionados por Marcello Damy de Souza Santos. A partir de 1939, Wataghin e seus assistentes começaram a divulgar os resultados sobre a produção múltipla e simultânea de partículas – os chuveiros penetrantes de mésons – fato científico confirmado mais tarde na Inglaterra. Publicaram diversos artigos na Physical Review e nos *Anais da Academia Brasileira de Ciências*. O prestígio alcançado resultou na colaboração com o físico americano Arthur Compton, para medir radiações cósmicas no interior paulista, por meio de balões de hidrogênio carregados com contadores Geiger. Os resultados alcançados pela Expedição Compton – o projeto da USP e da Universidade de Chicago — foram apresentados no primeiro evento internacional de física realizado no Brasil, o Symposium sobre Raios Cósmicos (Rio de Janeiro, 1941), promovido pela Academia Brasileira de Ciências.

Na mesma ocasião, Occhialini tentava pôr em funcionamento uma câmara de Wilson de grandes dimensões, telecomandada por contadores Geiger. Essa câmara era igual à de Patrick Blackett e, com esse equipamento, eles confirmaram, em 1933, a existência do pósitron e a produção de pares de elétrons resultantes de fótons de alta energia. O projeto da grande câmara de Wilson não deu certo em São Paulo, porém, alterou a direção da carreira de Cesar Lattes, permitindo que trocasse a física teórica, na qual trabalhava com Gleb Wataghin e Mario Schenberg [4], pela física experimental dos raios cósmicos sob a liderança de Occhialini, cujo magnetismo pessoal era contagiante e se transferia para muitos que o cercavam, despertando entusiasmo pela ciência, literatura e artes [5].

Wataghin e Occhialini, mesmo distantes dos centros produtores de ciência, não cortaram os vínculos com a elite européia e americana da física da qual faziam parte. A interlocução contínua possibilitou o acesso de brasileiros a laboratórios estrangeiros e, também, a formação de pesquisadores independentes e altamente qualificados. Na Universidade de São Paulo se fazia ciência de alcance internacional, mesmo estando instalada em um país da periferia.

Entretanto, o recrudescimento da Segunda Guerra Mundial inverteu a situação. Na Europa e nos Estados Unidos, os laboratórios de física voltaram-se para o desenvolvimento de tecnologia militar; em São Paulo, as pesquisas praticamente pararam. Por ser estrangeiro, Wataghin foi afastado da direção do Departamento de Física; a maioria dos físicos experimentais optou por fabricar secretamente artefatos para o Exército e a Marinha [6]. Gleb Wataghin, Cesar Lattes, e ocasionalmente Oscar Sala, bem como outros recém-formados, prosseguiram no estudo dos raios cósmicos.

Como temia ser deportado, depois de o Brasil ter declarado guerra aos países do eixo, Occhialini deixou a USP para permanecer incógnito como guia na floresta de Itatiaia (RJ). Depois, na cidade do Rio de Janeiro, trabalhou com Carlos Chagas Filho no Instituto de Biofísica, instituição de pesquisa vinculada à Universidade do Brasil. Retornou a São Paulo para dar um curso sobre raios-X, que habilitou Lattes a utilizar filmes fotográficos em física. Finalmente, o físico antifascista [7] deixou o país, disposto a participar da equipe do British Atomic Energy. Recusado por ser cidadão italiano, ficou numa situação difícil até conseguir ser contratado pela Universidade de Bristol – uma instituição privada mantida pela família Wills [8] – em junho de 1945. A indicação de Patrick Blackett fez com que fosse acolhido de braços abertos por Cecil Powell, pacifista e físico do H.H. Wills Physical Laboratory [9].

O encontro em Bristol

Cecil Powell desenvolvia um projeto de pesquisa de física nuclear, no qual utilizava placas fotográficas comuns – as mesmas que eram usadas para tirar retratos – para o estudo de reações nucleares nas emulsões. Dono de uma intuição especial e de incontestável habilidade técnica – em oposição ao seu desprezo pela erudição matemática e a uma certa apatia de Powell [10] – Occhialini logo percebeu a limitação do artefato. Assim, ambos teriam convencido C. Waller, químico e diretor da Ilford Ltd. – indústria inglesa de material fotográfico – a fabricar experimentalmente placas nas quais a densidade de prata da emulsão fosse superior à das placas usuais [11]. Foram necessárias várias tentativas para que a emulsão tivesse a densidade alterada e não tivesse aumentado o número de resíduos do fundo, uma vez que grãos indesejáveis mascaravam as trajetórias de prótons e de outras partículas em estudo. Em pouco tempo, a Ilford Ltd. começou a fabricar, em escala industrial, séries de placas de emulsão nuclear que se diferenciavam, em função do aumento da densidade e do aumento da sensibilidade [12].

Lattes ficou tão impressionado ao receber de Occhialini uma impressão positiva de fotomicrografia de traços de prótons e partículas α, obtidos com o novo concentrado de emulsão nuclear, que decidiu que iria para Bristol. Este detector era muito mais denso para ver mésons parando do que uma câmara de Wilson.

Em janeiro de 1946, depois de ter recebido uma fotografia de raio-γ tirada naquela câmara de Wilson, no porão do Departamento de Física da USP, Occhialini propôs a Powell convidar Lattes para integrar o grupo de Bristol. Occhialini prognosticou que o aperfeiçoamento da emulsão havia aumentado extraordinariamente as possibilidades de descobertas científicas. Mas a vantagem ficaria com o laboratório que realizasse o primeiro experimento com o poderoso artefato da Ilford e, para que isso fosse possível, o Wills Laboratory necessitava de "novas e jovens forças." Entretanto, dificilmente conseguiria adesões entre os físicos juniores recrutados para a guerra e que estavam sendo estimulados pelo governo a retornar à atividade acadêmica. Naqueles anos, os ingleses desprezavam a física experimental e as universidades do interior, pois o curso de graduação em física havia sido desviado para a formação de técnicos em eletrônica. Os atrativos do Departamento de Física de Bristol eram Neville Mott, físico teórico de renome internacional, e H. Frölich, apesar de Walter Heitler e Hans Bethe terem sido do quadro de professores na década de 1930. Essas circunstâncias levaram Occhialini a se lembrar de que Lattes lhe manifestara, quando deixou o Brasil, o desejo de fazer ciência no exterior [14].

Antes da Segunda Guerra Mundial, o Departamento de Física da Universidade de Bristol era basicamente financiado pelo Department of Scientific and Industrial Research – DSIR, pela Electrical Research Association e pelo Academic Assistance Council de ajuda aos refugiados. Depois da Guerra, quando também foi orientado para a pesquisa com fins militares, aumentaram os recursos financeiros externos para o Departamento. Provenientes do DSIR, Kodak, Iron & Steel Federation, Anglo-Iranian Oil, Electrical Research Association, Diamond Corporation e da Paul Fund [?] da Royal Society, os recursos possibilitaram a contratação de 13 físicos, entre 1946-50. Neste grupo estavam Giuseppe Occhialini (1945-48), Cesar Lattes (1946-48) e Ugo Camerini (1946-50). A Cesar Lattes foi oferecida uma bolsa de pesquisa, como complemento do salário de professor assistente de Física Teórica e Matemática que ele continuaria a receber da USP [15].

Lattes deixou, então, o porto do Rio de Janeiro num cargueiro, com passagem que o matemático Leopoldo Nachbin conseguira da Fundação Getulio Vargas (RJ). Foram 40 dias de viagem para chegar a um lugar isolado e frio, onde o *menu* do pós-guerra, no início de 1946, era pão e um prato de sopa. Lá, constatou que Powell e Occhialini continuavam utilizando as antigas chapas fotográficas no espalhamento nêutron-próton, numa pesquisa iniciada há mais tempo. Na sua avaliação, era difícil ver os traços de prótons na emulsão, quando a trajetória da partícula estivesse inclinada com relação ao plano da emulsão. Powell era conservador em ciência; em contrapartida, concedia aos subordinados ampla liberdade de trabalho, iniciativa e estímulo. Desse modo, podendo trabalhar no que mais lhe interessava e pelo tempo que lhe conviesse, Lattes abriu várias frentes. Ao contrário do que sucedia em São Paulo, onde era criticado pelo excesso de dedicação à física teórica, em Bristol ele mal dava conta das atividades experimentais [16].

Na época da sua chegada, as atividades do Wills Laboratory eram restritas à física nuclear, apesar de W. Heitler e G. Fertel terem trabalhado em raios cósmicos, expondo chapas fotográficas no observatório de Jungfraugh (3 000 m) [17]. Iniciou, então, uma pesquisa objetivando determinar a radioatividade do samário – na tabela periódica, um elemento terra rara que emite espontaneamente partículas-α e, segundo uma fraca evidência experimental, prótons – utilizando as novas placas de emulsão nuclear [18]. Ao mesmo tempo, desenvolvia a técnica da medida automática nas emulsões. Investia no futuro, ao procurar adquirir competência para aproveitar, mais tarde, naquilo que mais gostava: raios cósmicos. Calibrava intensamente as emulsões para estabelecer a relação entre o alcance, o comprimento do traço e a energia. Contatava o químico da Ilford Ltd. para resolver o problema do *fading* na imagem fotográfica. Lattes estava convencido de que seria possível modificar novamente as placas. Ugo Camerini, que trocou a USP por indicação de Lattes a Powell, participava da investigação. Lattes desenvolvia um método simples, mas revolucionário, para solucionar a questão, seguindo a orientação de Johnny Williamson, do Departamento de Química [19]. Na época da sua chegada, as atividades do Wills Laboratory eram restritas à física nuclear, apesar de W. Heitler e G. Fertel terem trabalhado em raios cósmicos, expondo chapas fotográficas no observatório de Jungfraugh (3 000 m) [17]. Iniciou, então, uma pesquisa objetivando determinar a radioatividade do samário – na tabela periódica, um elemento terra rara que emite espontaneamente partículas-α e, segundo uma fraca evidência experimental, prótons – utilizando as novas placas de emulsão nuclear [18]. Ao mesmo tempo, desenvolvia a técnica da medida automática nas emulsões. Investia no futuro, ao procurar adquirir competência para aproveitar, mais tarde, naquilo que mais gostava: raios cósmicos. Calibrava intensamente as emulsões para estabelecer a relação entre o alcance, o comprimento do traço e a energia. Contatava o químico da Ilford Ltd. para resolver o problema do *fading* na imagem fotográfica. Lattes estava convencido de que seria possível modificar novamente as placas. Ugo Camerini, que trocou a USP por indicação de Lattes a Powell, participava da investigação. Lattes desenvolvia um método simples, mas revolucionário, para solucionar a questão, seguindo a orientação de Johnny Williamson, do Departamento de Química [19].

Em Cambridge, noutra frente de trabalho, Lattes estudava reações provocadas pelo feixe de Em Cambridge, noutra frente de trabalho, Lattes estudava reações provocadas pelo feixe de dêuterons de 1 MeV do acelerador Cockroft-Walton, um gerador eletrostático de alta tensão. Estava interessado em *targets* leves – D, Li, Be, B, F – para esclarecer pontos obscuros relativos a essas reações. Insistia no objetivo de determinar a relação alcance-energia para partículas-α, prótons e dêuterons nas placas de emulsão concentrada, que se tornariam de uso generalizado na física nuclear e em raios cósmicos. Conforme comentou em carta ao amigo José Leite Lopes, os físicos nucleares, que antes desprezavam esse artefato, estavam se "preparando para uma verdadeira corrida"! [20] Cesar Lattes, Peter Cuer e Peter Fowler – os últimos alunos de graduação de Powell – conseguiram estabelecer a relação alcance-energia de prótons por meio da análise dos traços nessas chapas e, ainda, fizeram uma correção importante: a relação entre o alcance no ar e o alcance na emulsão não é constante [22]. Cesar Lattes investigava ainda com Pierre Cuer – estudante francês recém-chegado e interessado na técnica das emulsões – as desintegrações produzidas por nêutrons na mesma placa, carregando-a com sais de D, Li, Be.

Na parte teórica, estudava a ação das partículas nucleares nas emulsões fotográficas, em colaboração com Peter Burton do laboratório de pesquisa da Eastman Kodak e orientando de Mott. Ambos aventavam a possibilidade de somar esforços na fabricação de emulsões fotográficas sensíveis a elétrons para pesquisa sobre raios β, na fábrica da Kodak, em Londres [23].

Se a Segunda Guerra Mundial favoreceu o encontro de três físicos – Occhialini, Lattes e Powell – que preferiram prosseguir na produção de conhecimento científico a participar do esforço de guerra em seus países, a grande batalha para descobrir os mésons na natureza começou com o reencontro de Lattes e Occhialini. Juntos eles transformaram a vida no quarto andar da chamada *Cigarette Tour*, onde estava instalado o Wills Laboratory, ao implementarem as pesquisas em raios cósmicos [24]. Rompendo o marasmo em que se encontravam aqueles físicos e técnicos ingleses do grupo de Powell – complexados e frustrados por não terem se engajado nas atividades da guerra contra o nazi-fascismo, segundo Occhialini – estes reagiram à inquietude intelectual de ambos para mostrar ao mundo a contribuição civilizada da ciência [25].

O experimento no Pic du Midi

Obstinados em eliminar a ocorrência do *fading* na emulsão e em testar o método que estavam desenvolvendo, sem muito alarde, na detecção de partículas nos raios cósmicos, Lattes e Occhialini concluíram que precisavam reduzir o tempo de exposição das chapas. Ou seja, realizar experimentos em elevadas altitudes, onde a atmosfera terrestre é mais rarefeita e os raios cósmicos podem se chocar mais facilmente com os núcleos dos átomos da emulsão das chapas fotográficas especiais. E, como Occhialini iria passar nas férias pelos Pireneus, onde já expusera com Powell as placas fotográficas comuns, o Observatoire du Pic du Midi (2 800 m) tornou-se uma alternativa duplamente viável. O adido cultural do governo francês em Londres e o diretor desse observatório abriram as portas para o empreendimento. O físico Max Cosyns [26] estava entre os que garantiram a exposição no local de placas de emulsão C2 e B1 durante seis semanas. Todas estavam calibradas, mas eram diferentes entre si.

A diferença essencial era que só algumas emulsões estavam carregadas com bórax. A hipótese do método era de que a ação *antifading* do bórax manteria o poder de detecção das emulsões por mais tempo. Sem o bórax, ao contrário, as placas têm muito *fading* e perdem o poder de detecção em cerca de uma semana [27].

Em janeiro de 1947, na mesma noite em que retornou a Bristol, Occhialini revelou as placas e redigiu uma nota para a *Nature* [28] exaltando as vantagens das placas de emulsão nuclear para o estudo dos raios cósmicos. A hipótese era verdadeira: em altitudes elevadas a ação *antifading* do bórax permitiu registrar variedades de eventos com riqueza de detalhes, como acompanhar a trajetória de um lítio 8 – que Occhialini chamou de tarelo – emitindo uma partícula-β, decaindo em berílio 8. Este, por sua vez, emitia duas partículas-β que ele chamou de *martelo*. A visão proporcionada pelas placas com bórax mobilizou toda a força do laboratório para o estudo dos eventos normais de baixa energia dos raios cósmicos. Somada à autoridade de Occhialini, as discussões de Lattes, Camerini e Fowler eram essenciais ao processo da construção do méson. Dias depois, uma jovem microscopista, Marieta Kurz, encontrou um estranho evento: um "duplo méson" ou o que foi considerado o decaimento $\pi - \mu$ [30].

As medições de massa foram feitas em dois caminhos. Lattes, Camerini e Occhialini começaram refinando a técnica de contagem de grãos e do balanço de energia nos traços secundários das reações observadas. Muirhead, mais tarde auxiliado pelos outros pós-graduandos David King, Yves Goldschmidt-Clermont e David Ritson, enfrentaram o problema pela via do alcance-espalhamento múltiplo [31]. Lattes e Camerini enfatizaram que o total de energia produzida era maior que a massa do mésotron [32]. Depois de avaliar os resultados preliminares obtidos, consideraram que o méson pesado de Hideki Yukawa e o méson de Carl Anderson estavam ali [33]. Embora desconhecessem a teoria de Shoichi Sakata [34]

– de que haveria dois mésons na componente dura da radiação cósmica – tinham certeza de que se tratava de um processo fundamental. A institucionalização da física de partículas, até então subordinada à física nuclear, estava a um passo.

O segundo experimento nos raios cósmicos

O desafio era comprovar aos colegas de outros laboratórios que os mésons estavam de fato inscritos nas emulsões, e também, que eles não estavam sendo iludidos pelo artefato, isto é, as placas de emulsão nuclear carregadas com bórax deixadas aos raios cósmicos. Como poderia o grupo de Bristol garantir a interpretação do fenômeno, se o pequeno número de eventos analisados não permitia que se eliminassem estatisticamente os erros de cada um dos métodos utilizados para medições de massa?

Havia unanimidade quanto à necessidade de se conseguirem mais eventos, rapidamente, para que não fossem ultrapassados pelo Imperial College of Science and Technology. Temiam perder a batalha para D.H. Perkins, que expunha emulsões do tipo B1, sem bórax, em aviões da Real Força Aérea (9 000 m) [35]. Na disputa científica vence quem consegue publicar primeiro os resultados. Assim, desde que Lattes viu os primeiros resultados obtidos no Observatório do Pic du Midi, "(...) solicitou, quase exigindo, tomar parte em uma nova aventura" [36].

Prontamente, a Universidade de Bristol autorizou a sua viagem à Bolívia para repetir, nos Andes, o experimento dos Pireneus [37]. Teria sido fácil para Powell convencer o Prof. Tyndall, diretor do H.H. Wills Physical Laboratory, a liberar os recursos para o segundo experimento. Desde 1946, devido ao aprimoramento das placas Ilford, o Department of Scientific and Industrial Research – DSIR concedia substancial auxílio ao trabalho de Powell [38]. A descoberta de novas partículas significava maior compreensão das forças nucleares, possibilidades crescentes em termos de retorno político-militar para os países promotores e expectativas favoráveis para os acionistas de empresas envolvidas diretamente com a aplicação da ciência. Após a Segunda Guerra Mundial, a física ascendia como ciência-guia e a física nuclear era o assunto do dia. Neste cenário, o diretor oficializou a viagem de Lattes, numa cerimônia para as cerca de vinte pessoas que trabalhavam no Wills Laboratory. A partida, no dia 7 de abril de 1947, foi noticiada pelo *Bristol Evening World* [39].

Cesar Lattes desembarcou no Rio de Janeiro e atravessou o Brasil para pegar, em Corumbá, um avião rumo a Santa Cruz de la Sierra, Cochabamba, Oruro, Potosí e, finalmente, La Paz. Da capital da Bolívia, foi com o físico Ismael Escobar para Chacaltaya, a 20 km de La Paz. Na realidade, o estreito caminho de pedras entre desfiladeiros dava no que fora antes o Club Andino Boliviano e se transformara numa estação de meteorologia, onde havia uma minúscula e tosca instalação feita com quatro pedaços de madeira e duas plataformas pintadas de branco, formando um tronco de pirâmide. A 5,6 mil metros de altitude, sem se importar com a precariedade do lugar e esquecendo-se dos perigos da viagem, do frio e dos efeitos do oxigênio rarefeito a grandes altitudes, Lattes dispôs pequenas pilhas de placas na segunda plataforma, para receberem cerca de 100 mil vezes [40] mais partículas cósmicas do que aconteceria no observatório do Pic du Midi.

Um mês depois, ao retornar do Brasil, a altitude de Chacaltaya propiciou a Lattes observar, numa única placa revelada, dois duplos mésons completos, dois $\pi-\mu$. Mesmo manchada pela água turva com que foi revelada – na casa de Ismael Escobar – a emulsão permitiu-lhe ver que o alcance do traço secundário era de cerca de 600 micra. Pelo telégrafo, acertou com Powell que finalizaria o trabalho em Bristol. A ação *antifading* do bórax manteria aprisionados os eventos registrados nas placas, ao longo da viagem de regresso.

Na passagem pelo Rio de Janeiro, teve tempo para rever os amigos. Encontrou José Leite Lopes e Guido Beck, físico austríaco contratado professor visitante da Faculdade Nacional de Filosofia, e mostrou-lhes a placa de emulsão manchada. Juntos, no Departamento de Física e com o microscópio de Joaquim Costa Ribeiro [41], viram um terceiro duplo méson. Diante da incontestável evidência – a descoberta do méson-π e a desintegração $\pi-\mu$ – a emoção foi coletiva [42]. Em Bristol, mais de 30 duplos mésons foram observados e o singular tornou-se regular.

Mas faltava muito para o trabalho do laboratório estar concluído. Precisavam prosseguir as medições da massa, examinando detalhadamente cada evento. Muirhead continuaria pela via do alcance-espalhamento múltiplo, e Lattes pela contagem de grãos nos traços. O trabalho era árduo. Lattes, trabalhando compulsivamente, não estava comedido ao descrever as constatações para Leite Lopes, que estava envolvido com a teoria das forças nucleares. Na carta, que ele mesmo duvidava ser compreensível, fazia referências ao total e às características dos mésons capturados, aos mésons duplos, ao alcance e energia, à imprecisão da medida da massa, ao decaimento do μ, aos mésons positivos e negativos. Esquematizava para explicar, pois estava muito ansioso. Desenhava estrelas e elucidava que foram formadas a partir de mésons emitidos por núcleo em explosão. Alinhava as justificativas teóricas e argumentos provisórios, comparava com as considerações de Conversi, Pancini e Piccioni [43].

Placas fotográficas com densidade de prata especial, o bórax com que foram carregadas e a altitude de Chacaltaya permitiram confirmar a descoberta dos mésons-π naturais, produzidos no topo da atmosfera terrestre pelos raios cósmicos procedentes da galáxia. Mas a experiência, obviamente, não resultou de uma evidência natural, de uma observação empírica ou de uma mera dedução a partir das teorias dos mésons, e de ação isolada de um cientista. Múltiplos e inseparáveis fatores – sociais e cognitivos – perícia e competências administrativas, científicas e técnicas diversificadas, garantiram o sucesso. A montagem foi alicerçada na experiência de Occhialini em raios cósmicos, na perícia de C.

Waller da Ilford em aperfeiçoar o concentrado de emulsão, na capacidade de Lattes para testar um novo método de detecção e nas alianças bem estabelecidas, principalmente por Powell, com representantes de outras instituições: Ilford Ltd., Cavendish Laboratory, Cambridge High Tension, Embaixada da França em Londres, Observatoire du Pic du Midi, Servicio Meteorológico de Bolivia *etc.* [44].

A interpretação dos eventos perscrutados dependeu, no entanto, de pressupostos teórico-metodológicos. O experimento realizado permitiu a fixação de microscópicos pontos deixados na emulsão nuclear. Mas a constatação, isoladamente, dos eventos, não seria suficiente para entender o significado das partículas e das forças nucleares. Fenômenos naturais são passíveis de inúmeras definições que combinam conhecimento sistematizado, perícia, intuição, método, crenças, claras, ou mesmo obscuras, motivações. Trazidas ao H.H. Wills Physical Laboratory, as representações da natureza – obtidas nos Pireneus e Andes – são fruto de experiência profissional e de intenso trabalho em equipe. Elas significam o resultado da combinação entre a persistência de Powell, na utilização da emulsão fotográfica em física nuclear, as qualidades de Occhialini, na física de raios cósmicos, e a clareza de pensamento e dedicação de Lattes ao estudo dos mésons; dependeram de numerosas operações entre agentes de diferentes redes sociais, mobilizados por algum interesse comum; resultaram da consequente ação de intermediários de todos os gêneros: pessoas, objetos, natureza, sociedade, universidades, indústrias, governos, cientistas, técnicos, burocratas, professores, conhecimento e técnica. O processo foi contínuo, vencendo, paulatinamente, etapas para superar as dificuldades e controvérsias. E, nesse processo de construção de um fato científico, Cesar Lattes desempenhou um papel central [45].

O tempo da controvérsia

No plano do laboratório, as tarefas dos protagonistas da descoberta do méson estavam concluídas. A questão agora era persuadir os colegas de outras universidades sobre o que viram e sobre o que acreditavam ser. A natureza é pródiga e se deixa exprimir em discursos infinitos, que não são necessariamente nem coerentes, nem compatíveis entre si. E, de uma maneira geral, os outros físicos não estavam acreditando na descoberta. Não havia como alterar a tática. Na ciência, o caminho foi sempre o mesmo: discutir com colegas e publicar os resultados em periódicos importantes, para que os pares os leiam. Prioridade, originalidade e qualidade são alguns dos critérios de julgamento.

Até este ponto, o grupo de Bristol lutou contra o tempo, contra a concorrência de outro laboratório, contra dificuldades para representar a natureza e lutaram, de maneira especial, para interpretar as observações. Mas não tinha enfrentado o debate e a controvérsia entre os que acreditavam na existência de dois mésons e aqueles que acreditavam apenas no mésotron. O teste de resistência começou em Manchester, no final de junho, e na conferência sobre raios cósmicos e física nuclear, organizada por W. Heitler no Institute of Advanced Studies de Dublin, de 5 a 12 de julho de 1947. O trabalho apresentado por Powell – um mestre da palavra falada – justificava, interpretava e apresentava conclusões a partir dos experimentos realizados no Pic du Midi e Chacaltaya, comparando alguns resultados obtidos entre a exposição em altitudes elevadas e com a câmara de Wilson. O encontro de Dublin foi fundamental para que o enunciado da existência de dois mésons fosse reforçado, para que viesse a se tornar um fato científico. Hans Bethe e Robert Marshak emprestaram o manuscrito do artigo inédito sobre a hipótese de dois mésons, de um par de mésons [46]. Eugene Gardner fez o mesmo [47]. Isto foi importante porque, nos Estados Unidos, com poucas exceções, apenas se referiam aos mésotrons. Críticas e sugestões de físicos europeus – W. Heitler, L. Jánossy, H. Fröhlich, Christian Moller, Abraham Pais e outros – foram incorporadas antes da *Nature* publicar o trabalho de outubro. Os dois últimos consideravam a possibilidade de relacionamento entre tipos diferentes de partículas de massa intermediária [48].

No simpósio em Birmingham, no mesmo ano, os resultados foram reapresentados por Cesar Lattes, por determinação de Powell. Segundo suas declarações, foi difícil falar sobre a desintegração $\pi - \mu$, uma vez que muitos cientistas ainda se reportavam aos resultados de Pancini, Piccioni e Conversi acerca da vida média do méson-μ. O fato é que muitos físicos permaneciam incrédulos. Enquanto Lattes prosseguiu investigando intensamente a nova descoberta, refinando os cálculos, Powell prosseguiu viajando. Foi para Copenhague, Cracóvia e Suécia, de onde enviou as seguintes notícias para Lattes:

> "Apresentei dois seminários aqui sobre os nossos resultados e todos parecem concordar que eles constituem o mais importante avanço na física nos últimos anos. É muito compensador [?] que você tenha se unido tão fortemente e com uma substancial contribuição, especialmente porque você tem trabalhado tão intensamente e tem dado uma preciosa contribuição para o desenvolvimento dos nossos métodos" [49].

John Wheeler, presente aos eventos, constatou que as evidências do grupo de Bristol sobre os mésons pesados foram universalmente aceitas [50]. Lattes foi, ainda, à École Polytechnique de Paris e, em dezembro de 1947, apresentou um seminário na Sociedade de Física da Dinamarca e no Instituto de Física Teórica de Copenhague (atual Instituto de Física Niels Bohr), o grande centro de excelência em meados do século. Esteve também na residência de Niels Bohr. De lá, chegou até Lund, no Sul da Suécia, para seminário na universidade local [51].

O processo da descoberta do méson está registrado em seis trabalhos publicados na *Nature* e em dois artigos no periódico inglês *Proceedings of the Physical Society*, a revista da mais antiga sociedade de físicos (1874), apresentados no Quadro 1. Na época, era usual listar os autores em ordem alfabética. As exceções à regra ilustram a utilização de critérios de hierarquia [52].

Conclusão

Occhialini e Lattes não esperaram pela consagração dos resultados, que foi capitaneada, em 1950, por Powell [52]. Lattes surpreendeu aos de lá, inclusive Niels Bohr, com a decisão de deixar Bristol no melhor da festa. Com aparente impetuosidade, trocou o H.H. Wills Physical Laboratory pelo Radiation Laboratory de Berkeley, da Universidade da Califórnia, para tentar detectar mésons produzidos no ciclotron de 184", construído pelo físico Ernest Orlando Lawrence. Apesar do próprio Lattes ter tido dúvidas relacionadas à energia da máquina para a produção do méson π, a sua ida para Berkeley estava sendo preparada desde que passou no Brasil para ir à Bolívia, em abril de 1947.

Anônimos e desconhecidos nos raios cósmicos até 1947, os mésons eram fabricados no sincrociclotron do Radiation Laboratory of Berkeley sem que ninguém soubesse, de novembro de 1946 até o começo de fevereiro de 1948. Assim, foram detectados por Cesar Lattes e Ernest Gardner, quinze dias depois da chegada do físico brasileiro aos Estados Unidos com as emulsões nucleares da Ilford Ltd. [54].

Também em 1948, Occhialini foi para a Universidade Livre de Bruxelas, ressentido com Powell que assinou em primeiro lugar o trabalho *Nuclear Physics in Photographs* (Londres: Oxford Press, 1947). Lá, em colaboração com Connie Dilworth e Ron Paine, desenvolveu o método para o processamento de emulsões nucleares espessas, G5 da Ilford – mais sensível do que a C2 –, que possibilitou observar diretamente a completa desintegração $\pi - \mu$, o decaimento $\mu - e$ e, depois, outras partículas. Do mesmo modo, o grupo de Bruxelas fez estudos para as emulsões NT2 e NT4 da Kodak [55].

Análises históricas sobre a detecção do méson artificial e acerca do método para o desenvolvimento de emulsões mais sensíveis reforçam os motivos da indicação de Occhialini e Lattes para o Prêmio Nobel de Física, em 1950. Todavia, como a condecoração máxima da ciência foi entregue somente ao coordenador da pesquisa realizada em Bristol, Cecil Powell, e faltando ao historiador fontes documentais que não estão abertas à consulta na fundação Nobel, cabe salientar: a ideologia individualista do Prêmio Nobel, que continua alimentando a imagem errônea do cientista isolado em seu laboratório; as questões filosóficas e éticas que, na conjuntura da Guerra Fria, permearam o julgamento na comunidade de físicos, afastando da disputa aqueles que se envolveram direta ou indiretamente na fabricação da bomba atômica; as interferências geopolíticas na escolha de um colega. Nesta hipótese, hemisférios e países têm pesos distintos.

Artigos Publicados	Resumo
1. .G. Occhialini, C. Powell. Multiple disintegration process produced by cosmic rays. *Nature*, v. 159, p. 93-94, jan. 1947.	*"Carta ao Editor" destacando evidências obtidas no experimento de Pic du Midi e acentuando que as emulsões fotográficas especiais da Ilford C2 se prestam ao estudo dos raios cósmicos. Por meio de mosaicos de fotomicrografias, chama atenção para o artefato e resultados preliminares. Registra o papel de Lattes na preparação do material utilizado no experimento.*
2. .G. Occhialini, C. Powell. Nuclear disintegration produced by slow charged particles of small mass. *Nature*, v. 159, p. 186-190, fev. 1947.	*Artigo complementar à nota anterior, ilustrado com seis mosaicos de fotomicrografias, que reforça a qualidade das placas carregadas com bórax, a descrição das evidências observadas. Baseado em trabalho de Muirhead, sugere tratar-se de mésons negativos e positivos. Pela técnica de contagem de grãos, as partículas têm massa inferior à massa do próton. Registra a participação de Lattes e Fowler nas discussões.*
3. .C. Lattes, P. Fowler, P. Cuer. A study of a nuclear transmutations of light elements by the photographic method. *Proceedings of the Physical Society*, v. 59, p. 883-900, fev. 1947.	*Análise de reações com alvos leves (D, Li, Be, B, F) provocadas por dêuterons de 900 KeV produzidos pelo acelerador de Cambridge. Informa que foram utilizadas as placas do tipo B1 para registrar os traços de desintegração de partículas. As características da emulsão são também examinadas, em particular a relação alcance-energia.*
4. .C. Lattes, P. Fowler, P. Cuer. Range-energy relation for protons and α-particles in the new Ilford "Nuclear Research" Emulsions. *Nature*, v. 159, p. 301-302, mar. 1947.	*Pequena nota sobre os resultados descritos no trabalho acima, visando determinar a relação alcance-energia de partículas.*
5. .C. Lattes, G. Occhialini. Determination of the energy and momentum of fast neutrons in cosmic rays. *Nature*, v. 159, p. 331-332, mar. 1947.	*"Carta ao Editor" detalhando experimentos em que utilizaram emulsões C2 e B1, e comparando os resultados obtidos com o feixe de dêuterons do acelerador de Cambridge e nos raios cósmicos, em Pic du Midi. Enfatiza as vantagens das placas carregadas com bórax para a determinação da energia e do momento na desintegração do méson.*
6. .C. Lattes, H. Muirhead, G. Occhialini, C. Powell. Process involving charged mesons. *Nature*, v. 159, p. 694-697, maio 1947.	*Ressaltando a conveniência de aplicar o termo méson para a partícula com massa intermediária entre o elétron e o próton, o artigo apresenta os primeiros casos observados da desintegração do π-μ. Menciona a ocorrência de dois outros eventos nucleares produzidos por méson, similares aos observados por Perkins e por Occhialini e Powell (Quadro 3, referências 28 e 30). Os métodos utilizados, as dificuldades de medição da massa do méson e as conclusões são descritas.*
7. .C. Lattes, G. Occhialini, C. Powell. Observations on the tracks of slow mesons in photographic emulsions. *Nature*, v. 160, p. 453-456, 486-492, out. 1947.	*Trabalho de impacto do grupo, pois analisa 31 eventos, comparando as massas das partículas primária e secundária na desintegração do méson. As duas partículas observadas recebem o nome de méson-π e méson-μ, com massas intermediárias entre elétrons e prótons, uma desintegrando da outra com a participação de uma partícula neutra, não observada.*
8. .C. Lattes, G. Occhialini, C. Powell. A determination of the ratio of the masses of π and μ mesons by the method of grain-counting. *Proceedings of the Physical Society*, v. 61 (1948), 173-183.	*Análise das medidas de massa pela contagem de grãos dos mésons-π e μ para reforçar as conclusões do trabalho anterior. Com informação sobre cada um dos 31 eventos observados, possibilita comparações minuciosas entre os dois mésons. O método de Perkins foi também utilizado para a determinação da massa das partículas. Além de tabelas e gráficos demonstrativos dos resultados, há sete exemplos de fotomicrografia do decaimento π-μ.*

Notas e Referências Bibliográficas

[1] Cesar Lattes (Brasil, 1924): estudante de física da USP (1939-43); professor da USP (1944-49; 1960-67); research associate do H.H. Wills Physical Laboratory da Universidade de Bristol (1946-47); expert consultant do Radiation Laboratory of Berkeley da Universidade da Califórnia (1948-49); fundador e diretor científico do Centro Brasileiro de Pesquisas Físicas (1949-55), bem como pesquisador emérito; professor catedrático da Universidade do Brasil (1949-67); fundador e conselheiro do CNPq (1951-55); research associate do Institute for Nuclear Studies da Universidade de Chicago (1955-56); do College of Science Literature and Arts da Universidade de Minesota (1956-57); professor-conferencista da Universidade de Pisa (1964-65); professor da Unicamp (1967-84).

[2] Giuseppe Occhialini (Itália, 1907-93): graduado em física pela Universidade de Florença (1929); pesquisador assistente do laboratório de Arcetri da mesma universidade (1930-37) e bolsista do Cavendish Laboratory de Cambridge (1931-34); professor da USP (1938-42, 1944); pesquisador do Instituto de Biofísica do Rio de Janeiro (1943); pesquisador do H.H. Wills Physical Laboratory da Universidade de Bristol (1945-48); Universidade Livre de Bruxelas (1948-52); Universidade de Gênova (1950-52); pesquisador visitante do Centro Brasileiro de Pesquisas Físicas (1952); professor da Universidade de Milão (1952-93); Massachusetts Institute of Technology (1960); conselheiro do European Space Research Organization. Seu pai, Augusto Occhialini, também era físico e trabalhou no laboratório de física de Arcetri e no Wills Laboratory.

[3] Ver: A. Videira, M.C. Bustamante. Gleb Wataghin en la Universidad de São Paulo: un momento culminante de la ciencia brasileña. *Quipu*, n. 10, p. 263-284, 1993. Sobre os precursores da física no Brasil, ver: J.M. Bassalo, José Maria Filardo. *Crônicas da Física*. Belém: Universidade Federal do Pará, 1987. t. 1, 1990. t. 2, 1991. t. 3, 1994. t. 4; F. Caruso, A. Troper (ed.). Perfis. Rio de Janeiro: CBPF, 1997. A. Marques (ed.). *Cesar Lattes 70 anos: a nova física brasileira*. Rio de Janeiro: CBPF, 1994.

[4] C. Lattes, G. Wataghin. Estatística de partículas e nucleons e sua relação com o problema de abundância dos elementos. *Anais da Academia Brasileira de Ciências*, v. 17, n. 4, p. 269, dez. 1945; C. Lattes, G. Wataghin. On the abundance of nuclei in the universe. *Physical Review*. v. 69, p. 237, fev. 1946; C. Lattes, M. Schenberg, W. Schutzer. Classical theory of charged point-particles with dipole moments. *Anais da Academia Brasileira de Ciências*. v. 19, n. 3, p. 193-245, set. 1947.

[5] A. Marques. *op. cit.* nota 3; C. Lattes. *O nascimento das partículas elementares*. Palestra proferida na abertura da Comemoração dos 45 anos do CNPq em 11 out. 1996. Editada por F. Caruso & A. Troper, Ciência e Sociedade, CBPF-CS-023/97, Rio de Janeiro, CBPF, 1997; a ser publicado em inglês no livro *Wonders and Frontiers of Science*, pp. 173-189, Rio de Janeiro, IMPA. V. Telegdi. N.E.: *Cf.* Capítulo 12 deste volume. *Giuseppe Occhialini. Commemoration in Florence*. Florença, 11 de setembro de 1995. Há informações sobre Occhialini no site da Agenzia Spaziale Italiana.

[6] M.D. Santos. Homenagem a Cesar Lattes. In: Marques, Alfredo (ed.). *Cesar Lattes 70 anos: a nova física brasileira*. Rio de Janeiro: CBPF, 1994, p. 57; M.D.S. Santos. Marcello Damy. Entrevista concedida a Eduardo Fernandes. São Paulo: IPEN/USP, 1994. p. 31-37.

[7] Colegas brasileiros garantem que Occhialini nunca fez apologia do regime de Mussolini, como outros professores italianos contratados pela USP e

Universidade do Brasil. Já o ítalo-russo Gleb Wataghin foi simpatizante do fascismo.

[8] Segundo Lattes, a família era proprietária de indústria de cigarros.

[9] O H.H. Wills Physical Laboratory foi fundado em 1927; inicialmente, denominava-se Henry Herbert Wills Physics Laboratory. Tyndall, A.M. *A history of the department of physics in Bristol*, 1876-1948. Bristol: 1956. [mss, Bristol Library].

[10] Cecil Powell (Inglaterra, 1903-1969) foi orientando de C.T.R. Wilson, no Cavendish Laboratory. Em 1928, foi contratado assistente de pesquisa da Universidade de Bristol, pelo próprio diretor do Wills Laboratory. idem, p. 25. Segundo Rosemary Fowler (Entrevista concedida a Cássio Vieira. Bristol, mar. 1997), recém-formado, ele teria morado na América Central.

[11] Tyndall, *op. cit.* nota 9, p. 29 informa que, nos anos 30, já haviam percebido a necessidade de um novo tipo de emulsão e que Powell procurou C. Waller. Lattes (*Ciência Hoje*, v. 19, n. 112, p. 10-22, ago. 1995. p. 13) afirma que a iniciativa foi de Occhialini. Em notas biográficas de Occhialini, porém, consta que o empenho foi de Powell e Occhialini.

[12] Também há divergência nos depoimentos dos contemporâneos acerca do número de vezes que a densidade de prata das placas foi aumentada, em 1945-46. Lattes, idem, p. 13; H. Muirhead, Encounters with Giulio Lattes. In: Bellandi Filho, Chinellato, Pemmaraju (ed.). *Topics on cosmic rays. 60th anniversary of C.M.G. Lattes*. Campinas: Unicamp, 1984. v. 1, p. 14; e Perkins, pp. 17-23.

[13] O Departamento de Física de Bristol, notadamente, empenhou-se no desenvolvimento da tecnologia do rádio.

[14] G. Occhialini. Cesar Lattes: the Bristol years. In: Bellandi Filho, Chinellato, Pemmaraju (ed.). *Topics on cosmic rays*. 60th anniversary of C.M.G. Lattes. Campinas: Unicamp, 1984. v. 1. p. 6; Muirhead, *op. cit.* nota 12, p. 14; Lattes, entrevista à autora em 1996.

[15] Tyndall, *op. cit.* nota 9. p. 31-32.

[16] Carta de Cesar Lattes a Leite Lopes em 21 abril de 1946 (Arquivo Leite Lopes); Lattes, *op. cit.* nota 11, p. 14.

[17] Tyndall, *op. cit.* nota 9, p. 29; Occhialini, *op. cit.* nota 14, p. 7.

[18] C. Lattes, P. Cuer. Radioactivity of samarium. *Nature*. v. 158, p. 197-198, ago. 1946.

[19] Muirhead, *op. cit.*, nota 12, p. 14.

[20] Carta de Cesar Lattes a Leite Lopes em 21 de abril de 1946 (Arquivo Leite Lopes).

[21] Peter Howard Fowler – neto de Ernest Rutherford – ingressou no curso de física da Universidade de Bristol, em 1940. Devido à Guerra, interrompeu o curso entre 1942-45. Seu pai também era físico.

[22] C. Lattes, E. Samuel, P. Cuer. Radioatividade do samário, utilização da placa fotográfica para a determinação de baixas concentrações de material radioativo. *Anais da Academia Brasileira de Ciências*, v. 19, p. 1-15, fev. 1947; C. Lattes, P. Fowler, P. Cuer. Range-energy relation for protons and α-particles in the new Ilford Nuclear Research Emulsions. *Nature*. v. 159, p. 301-302, mar. 1947.

[23] Carta de Cesar Lattes a Leite Lopes em 21 de abril de 1946 (Arquivo Leite Lopes).

[24] M. Schenberg. Cesar Lattes, grande físico e personalidade extraordinária. In: Bellandi Filho, Chinellato, Pemmaraju (ed.). *Topics on cosmic rays. 60th anniversary of C.M.G. Lattes*. Campinas: Unicamp, 1984. v. 1. p. 11. Occhialini, *op. cit.* nota 14, p. 6.

[25] *idem*, p. 7.

[26] Max Cosyns era professor da Universidade Livre de Bruxelas (1945-53). Outros nomes constam dos agradecimentos nos artigos 1, 2 e 5 relacionados no Quadro 1.

[27] C. Lattes. My work in meson physics with nuclear emulsion. In: Bellandi Filho, Chinellato, Pemmaraju (ed.). *Topics on cosmic rays. 60th anniversary of C.M.G. Lattes*. Campinas: Unicamp, 1984. v. 1, p. 2; Muirhead, *op. cit.* nota 12, p. 14.

[28] Quadro 1, artigos 1 e 2.

[29] Muirhead, *op. cit.* nota 12, p. 15 comentou que um desses dois brasileiros foi quem lhe falou, pela primeira vez, sobre Paul Dirac. Na biblioteca de Bristol faltavam as obras de W. Heitler, G. Gamow, P. Dirac, dentre outras, imprescindíveis ao trabalho. Lattes atribuiu o fato à guerra. Carta de Cesar Lattes a Leite Lopes em 21 abril de 1946 (Arquivo Leite Lopes).

[30] Lattes, *op. cit.* nota 27, p. 2-3. Atualmente, o μ não é considerado um méson, é denominado lépton-μ, e o méson-π não é mais considerado o único agente das forças nucleares. A. Marques. *A descoberta do méson-π*. Rio de Janeiro: CBPF, 1998. [mss]

[31] *idem*, p. 2; Muirhead, *op. cit.* nota 12; Lattes, *op. cit.* nota 11, p. 14; Arquivo Leite Lopes: correspondência com Cesar Lattes.

[32] Occhialini, *op. cit.* nota 14, p. 8.

[33] Em 1935, Hideki Yukawa (Prêmio Nobel, 1936, da Universidade de Kioto) previu a existência de uma partícula de carga igual e massa 300 vezes maior do que à do elétron. Por ter massa intermediária entre as partículas elementares conhecidas, foi chamada de méson, em grego, do *meio*. Em 1937, os americanos Seth Henry Neddermeyer e Carl David Anderson (Prêmio Nobel, 1936) mostraram que parte da componente penetrante existente na radiação cósmica possuía partículas com massas intermediárias entre as do elétron e as do próton. Tornaram-se conhecidas como mésons (abreviatura da denominação original, mésotron), com 200 m_e, segundo J.C. Street e E.C. Stevenson. Ver F. Close, M. Marten, C. Sutton. The *particle explosion*. Nova York: Oxford University Press, 1987; Marques, *op. cit.* nota 3, p. 33; Quadro 1, artigo 6.

[34] A previsão de Shoichi Sakata (1942) não ficou conhecida, porque a Segunda Guerra Mundial impediu a circulação de artigos publicados no Japão. Ver os comentários de Fujimoto, pp. 25-32. Marques, *op. cit.* nota 3, p. 33. Veja o depoimento de Perkins, pp. 17-23.

[35] Marques, *op. cit.* nota 3, p. 33. Veja o depoimento de Perkins, pp. 17-23.

[36] Occhialini, *op. cit.* nota 14, p. 8.

[37] Carta de Cesar Lattes a Gleb Wataghin de 21 de março 1946 [1947?] comunicando detalhes da viagem. Fac-símile em A. Hamburguer, A.A. Videira. *Os cinqüenta anos do méson-π*. São Paulo: USP, 1997. p. 26. O documento sugere que o grupo de Bristol havia feito experimento anterior na Bolívia, pois Lattes foi também encarregado de processar as placas que estavam lá "desde junho", neste caso, de 1946.

[38] Tyndall, *op. cit.* nota 9, p. 32.

[39] *Bristol Evening World*, 9 abril 1947.

[40] Lattes, informação à autora em 1996.

[41] Joaquim Costa Ribeiro era o chefe do Departamento de Física da Faculdade Nacional de Filosofia.

[42] J. Leite Lopes. Cesar Lattes, o Centro Brasileiro de Pesquisas Físicas e a nova física no Brasil. In: A. Marques (ed.). Cesar Lattes 70 anos: a nova física brasileira. Rio de Janeiro: CBPF, 1994. p. 80; C. Lattes. Leite Lopes and physics in Brazil: a personal testimony. In: N. Fleury, S. Joffily, J.M. Simões, A. Troper (ed.). *Leite Lopes Festschrift*. Singapura: Word Scientific, 1988. p. 4; Lattes, *op. cit.* nota 11, p. 15.

[43] Conversi, Pancini, Piccioni. Physical Review. v. 68, 1945; Conversi, Pancini, Piccioni. *Physical Review*. v. 71, 1947; Carta de Lattes a Leite Lopes em 16 de julho de 1947 (Arquivo Leite Lopes).

[44] Um indicador desses vínculos são as notas de agradecimento que constam nos artigos.

[45] Occhialini, *op. cit.* nota 14, p. 7.

[46] Bethe, Marshak. *Physical Review*. v. 72, 1947.

[47] Gardner. *Physical Review*. v. 72, 1947. No ano seguinte, Lattes irá trabalhar ao lado de Gardner no Radiation Laboratory of Berkeley.

[48] Quadro 1, artigo 7.

[49] Carta de Powell a Lattes em 1947 (Arquivo Leite Lopes). Fac-símile em Leite Lopes, *op. cit.* nota 42, p. 80.

[50] Carta de J. Weeler a Leite Lopes em 3 nov. 1947 (Arquivo Leite Lopes).

[51] Lattes, *op. cit.* nota 11, p. 15; Bellandi Filho, Chinellato, Pemmaraju, *op. cit.*, v. 2, anexo.

[52] Quadro 1, artigos 3 e 4. Os co-autores de Lattes eram estudantes de graduação.

[53] Cecil Powell obteve o Prêmio Nobel de 1950, pelo desenvolvimento do método fotográfico de estudo do processo nuclear e pelas descobertas feitas observando mésons com esse método. Ele soube dar visibilidade ao seu trabalho, iniciado nos anos 1940, com emulsões fotográficas, explorar as potencialidades de seu grupo, reaproveitar os resultados experimentais e, ainda, tinha o dom de convencer plateias. Trocou intensa correspondência com Niels Bohr e, em 1948, organizou com N. Mott – diretor do Wills Laboratory – um simpósio internacional sobre radiação cósmica em Bristol. Lattes e Gardner não enviaram o trabalho sobre a produção artificial de mésons. Lattes, assim como Niels Bohr, foram ao simpósio internacional de física nuclear em Cambridge (Inglaterra), realizado na mesma época. Carta de Powell para Niels Bohr em 31 ago. 1948. Bohr Scient. Corr. (Niels Bohr Archive).

[54] Gardner, Lattes. Production of mesons by the 184-inch Berckeley cyclotron. *Science*, n. 107, p. 270-271, 1948; J. Burfening, J. Gardner, E. Lattes, C. Positive mesons produced by the 184-inch Berkeley cyclotron. *Physical Review*. v. 75, n. 53, p. 382-387. fev. 1949. Cesar Lattes recebeu o Prêmio de Física da Academia de Ciências do Terceiro Mundo em 1987; o Prêmio Bernardo Houssay, de 1978, da Organização dos Estados Americanos; e, no Brasil, o Prêmio Einstein, de 1950, da Academia Brasileira de Ciências, o *Prêmio Ernesto Lopes da Fonseca Costa*, do CNPq, dentre muitas distinções recebidas

também na Bolívia e Venezuela. É professor honoris causa da Universidade de São Paulo.

[55] Occhialini recebeu diversos prêmios e distinções na Itália e era doutor *honoris causa* da Universidade de Bristol e da Universidade de Bruxelas.

[56] A.M.R. Andrade. Físicos, mésons e política: a dinâmica da ciência na sociedade. São Paulo e Rio de Janeiro: Hucitec, Mast. 1998. Há diversas notas sobre a contribuição de Occhialini – de autoria de V. Telegdi, Peyrou, L.A. Antonelli, dentre outros – reunidas no departamento de física da Universidade de Roma La Sapienza, bem como na Universidade de Milão.

14

Os Físicos e suas Histórias: os Usos e as Versões da Detecção do Méson PI

Antonio Augusto Passos Videira & Cássio Leite Vieira

"A descoberta e análise do **decaimento píon-múon** é parte da história da ciência. A observação da produção artificial de píons, quando Lattes deixou Bristol indo para Berkeley, pertence a uma outra época. Se ele só tivesse trabalhado aqueles dois anos em Bristol, já mereceria um lugar na história."

Giuseppe Occhialini

("Cesar Lattes, os anos em Bristol")

Os cientistas e suas versões

Cinquenta anos depois da detecção do méson pi (1947), uma circular da Universidade de Bristol, comunicando a realização de uma conferência internacional comemorativa do evento, afirmava o seguinte:

"This conference forms part of our celebrations to mark the 50th anniversary of the discovery of the pion *by Nobel Laureate Cecil Powell.*" (grifos nossos).

É óbvio que não se pode esperar que o texto sucinto de uma circular relate com precisão e detalhes a história de uma descoberta relevante para a física deste século. O anúncio serve, contudo, como exemplo de uma tese considerada hoje trivial: há várias maneiras de se contar a mesma "história", que pode ser transformada segundo a perspectiva de quem se propõe a contá-la.

Os cientistas sempre perceberam a importância de apresentar suas próprias versões (ou interpretações) dos fatos históricos em que tomaram parte. A prática da elaboração de suas versões remonta aos tempos da fundação das primeiras academias científicas no final do século XVII. Já naquela época, os cientistas reconheciam a necessidade de firmar a ciência como conhecimento verdadeiro e socialmente relevante – a teologia e a filosofia ainda eram sérios concorrentes à sua pretensão de ser o melhor modo de explicar o mundo natural.

Para realizar esse objetivo, os cientistas criaram academias, usando-as como locais para a apresentação e a discussão públicas de suas ideias, bem como das estratégias para a obtenção de reconhecimento junto aos governos e à sociedade. Assim sendo, os cientistas precisavam determinar (o mais inequivocamente possível) e difundir o sentido de suas teorias e de seus feitos. No contexto deste trabalho, interessa-nos assinalar a conscientização mostrada pelos cientistas acerca dos rumos que suas descobertas poderiam tomar, isto é, como elas deveriam ser compreendidas e incorporadas por pares que delas não tomaram parte ou por aqueles que não praticam a ciência.

A narração histórica sobre a ciência raramente aparece muito tempo depois do acontecimento. Podemos supor que essa necessidade de elaboração e publicação das versões produzidas pelos cientistas decorre da hipótese, já mencionada, de que eles reconhecem o quão importante é poder contar com apoio dos não-cientistas, apoio este passível de ser transformado em reconhecimento social e em recursos financeiros. No processo de elaboração de suas versões, os cientistas avaliam (positiva ou negativamente) o papel e a relevância de suas descobertas e das de seus pares.

Através da história que elaboram, é reforçada uma hierarquia científica (nela estabelecendo-se um *ranking* de competência intra-pares) a qual chega, inclusive, a ser considerada como natural e óbvia. Essa naturalidade e obviedade originam-se na maneira segundo a qual os cientistas contam suas vidas profissionais, suas formações, seus sucessos e fracassos.

Apesar da importância que deve ser creditada às versões históricas contadas pelos cientistas, é imprescindível ter em mente que elas não são neutras, ou seja, nelas encontramos juízos de valor, preferências políticas, opções filosóficas e relações pessoais de amizade. Todos esses fatores desempenham papel preponderante na elaboração dessas narrações. Nem sempre é simples e imediato reconhecer a presença de todos esses fatores *subjetivos* e como eles influenciam a elaboração das versões.

Para aumentar essa dificuldade, concorre o fato de que os cientistas pretendem fazer crer em seus discursos que a ciência é neutra e objetiva. Em outras palavras, os fatores *subjetivos* acima expostos, em princípio, não desempenhariam papel significativo em suas narrações históricas. Objetivas ou subjetivas, neutras ou não, as declarações dos cientistas a respeito do desenvolvimento da ciência constituem material imprescindível para o historiador da ciência profissional.

Em países pouco desenvolvidos cientificamente, a avaliação madura e relativamente isenta da importância de seus pesquisadores pode ser influenciada pela necessidade de esses países disporem de cientistas de renome internacional, isto é, aceitos e integrados aos centros de excelência científica mundiais. A presença de cientistas de nível internacional em suas universidades e seus centros de pesquisa significaria, entre outras coisas, que esses países já teriam atingido um razoável grau de civilização e desenvolvimento.

Portanto, difundir o processo de evolução histórica das descobertas e das teorias científicas e influenciar os seus modos de recepção não são atividades estranhas à própria ciência. Um exemplo interessante do que estamos afirmando diz respeito à participação de físico brasileiro nos eventos que determinaram a comprovação experimental da existência do méson pi. Como veremos ao longo deste artigo, esses eventos foram contados de modos diferentes, variando de acordo com a perspectiva daquele que os narra. Acreditamos que a versão "brasileira" sobre a detecção do méson pi foi elaborada e difundida em função do fato de um grupo de cientistas brasileiros ter percebido na época que ela poderia ser um elemento fundamental para a mudança qualitativa que eles queriam implementar na ciência que aqui se fazia.

A versão de Powell

Uma vez que os cientistas não se furtam a apresentar versões para os eventos nos quais tomaram parte, vejamos o que nos é contado pelo prêmio Nobel de física de 1950, o físico inglês Cecil Frank Powell (1903-1969). Em sua autobiografia – a qual Powell não chegou a finalizar em vida, mas que foi publicada postumamente –, ele apresenta sua versão para a descoberta do méson pi. O que se destaca nela é o estilo adotado, diferente daquele frequentemente encontrado numa obra desse gênero.

No capítulo 5, dedicado ao méson pi, parece-nos que a escolha de Powell recai sobre o papel que as técnicas das emulsões nucleares tiveram em seus trabalhos, não nos dando detalhes sobre a evolução histórica de como esses aperfeiçoamentos foram obtidos – esse detalhamento é deixado para o livro que escreveu com D. Perkins e P. Fowler (ver referências). A linguagem de sua autobiografia aproxima-se do estilo de redação próprio ao que encontramos em textos científicos. Mas, para o que se segue, importa-nos observar que ele reconhece que, sem as novas chapas de emulsão nuclear da Illford, não teria sido possível comprovar experimentalmente a existência do méson pi:

"There were a number of ways in which it seemed possible that the recording properties of emulsions could be improved: by increasing the size and sensitivity of individual grains, for example, or by increasing the number of grains in unity volume of emulsion. C. Waller was the research chemist at Ilford's at that time and Ilford's method of manufacture were such that he found it possible to make emulsions with a very substantial increase in the concentration of silver bromide. When the new emulsions were exposed and developed it was clear that a very remarkable improvement in recording properties has been achieved." [1].

Powell continua seu relato afirmando que, uma vez conseguido o aperfeiçoamento das novas chapas, o físico italiano Giuseppe Occhialini (1907-1993) partiu para o Pic du Midi, nos Pireneus franceses, e lá as expôs. De volta a Bristol e reveladas as chapas, seus conteúdos apresentaram traços de eventos produzidos por partículas de raios cósmicos de alta energia: "It was as if, suddenly, we had broken into a walled orchard, where protected trees had flourished and all kinds of exotic fruits had ripened in great profusion." [2].

Foi com essas chapas que o grupo de Bristol constatou pela primeira vez a existência do méson pi.

Fato interessante é que o próprio Powell fornece elementos que se contrapõem à versão apresentada em sua autobiografia. A versão de Powell pode ser, por exemplo, "contestada" por um trecho de uma carta que Lattes enviou a Leite Lopes ainda em 1947. Nela, Lattes reproduz, de próprio punho, uma declaração de seu então chefe. Na carta a Leite Lopes, lê-se o seguinte:

"I have given two lectures here [Copenhague] on our results and everybody seems to agree that they constituted the most important advance in physics in the past few years. It is very satisfactory [palavras ilegíveis] *that you have been so closely associated with a substantial contribution, specially because you have worked so hard and have made such a valuable contribution to the development of our method (....)*" [3] (grifos nossos).

Lendo essas palavras, podemos indagar as razões que fizeram com que Powell, cerca de 20 anos depois, não mencionasse o nome de Lattes em seu texto autobiográfico. O conhecido perfil político e humano de Powell faz com que descartemos a hipótese de ele ter agido com má-fé (ver ref. [4]). Não querendo nos estender sobre este ponto, preferimos observar que é provável que sua concepção de história ou mesmo o estilo que adotou em sua autobiografia sejam os possíveis responsáveis por essa omissão.

Os ecos de Powell em Lock

A versão de Powell é aceita e incorporada por um de seus ex-colaboradores, Owen Lock, quem, em um artigo comemorativo dos 50 anos da descoberta do méson pi, afirma, entre outras coisas, que o seu antigo chefe foi o principal responsável pelo aperfeiçoamento da técnica que tornou possível a confirmação da existência do píon.

A menção indireta – (···) *and his collaborators* – aos outros cientistas e técnicos envolvidos nos fatos aponta para a ideia que Lock tem da história da ciência: só vale mencionar os grandes homens e os principais fatos:

"None of the scientists at the conference [Conferência de Shelter Island, junho de 1947] knew that two-meson decay events have already been observed some weeks earlier by *Cecil Powell and his collaborators* in Bristol, using the little known photographic emulsion technique, *but which in Powell's hands became a powerful research tool.*" [5] (grifos nossos).

Como mostram as declarações de Powell e Lock, eles sabiam que foi o aperfeiçoamento das emulsões nucleares e a procura de locais adequados para as exposições à radiação cósmica que possibilitaram a descoberta que celebrizou mundialmente o grupo de Bristol.

No entanto, foi preciso que o grupo de Powell vencesse uma desconfiança latente e que permeava a comunidade dos físicos quanto ao uso do método fotográfico para o estudo dos raios cósmicos. Podemos supor, então, que os físicos que "habitavam" o quarto andar da *Cigarette Tower* apostaram em uma técnica que na época era pouco confiável. Vejamos o que Powell e seus colaboradores mais próximos da década de 1950 (D. Perkins e P. Fowler) escreveram sobre esse ponto:

"By 1945, a modest field of activity for the photographic method had been found, but it was generally believed that any contributions which could be made with it were restricted in scope, and that it was the secondary importance comparison with counters and expansion chambers; see, for example Smyth (1946)." [6].

Como eles afirmam, a descoberta do méson pi foi o fato mais marcante para que as resistências a esse método pudessem ser superadas. A aposta havia se mostrado correta:

"The discovery of the pi-meson lead to general recognition of the value of the photographic method, and to its widespread use of study of nuclear physics and the cosmic radiation."[7].

A versão de Occhialini

Há 50 anos, uma oportunidade para o reconhecimento da comunidade científica brasileira surgiu quando o jovem físico paranaense Cesare Mansueto Giulio Lattes (1924-2005) participou da detecção natural do méson pi e – fato ainda mais marcante – da confirmação de sua produção no acelerador de Berkeley (EUA), esta última em colaboração com o físico norte-americano Eugene Gardner (1913-1950).

Mesmo não sendo preocupação de Lattes se tornar conhecido publicamente, sua participação nesses eventos foi amplamente divulgada pela imprensa no Brasil e no exterior. Os jornais e as revistas brasileiros da época citaram a descoberta como um exemplo das potencialidades da ciência brasileira (ver ref. [8]).

Retornando à autobiografia de Powell, é interessante notar que o físico inglês não cita o nome de Lattes quando se refere à descoberta do méson pi, partícula que confirmou a teoria do físico japonês Hideki Yukawa (1907-1981) sobre a interação nuclear forte.

Numa homenagem aos 60 anos de Lattes, Occhialini escreveu, em artigo publicado na *Folha de São Paulo* (21/7/1984) e reproduzido em *Ciência e Cultura* (ver ref. [9]), que o físico brasileiro havia desempenhado papel importante nos eventos ocorridos em 1947:

"Lattes chegou e a vida no quarto andar do Royal Fort se transformou. (···) Quando ele chegou a Bristol a atividade do quarto andar era restrita à física nuclear. Ele trabalhou na desintegração do samário e nas experiências de espalhamento nuclear realizadas por Powell, mas seu coração se fixou nos raios cósmicos.

"Nesse meio tempo ele dominou a nova técnica. Com Peter Fowler, então um estudante de pós-graduação, ele estabeleceu a relação alcance-energia nas novas emulsões concentradas. *Após sua descoberta de que mesmo nas novas emulsões o desaparecimento gradual da imagem era importante*, ficou claro que o tempo de exposição das chapas que eu expus no Pic du Midi deveria ser encurtado. O ganho de algumas semanas causado por isso mostrou-se crucial. (···) Sua contribuição foi muito importante. Ele trouxe não somente sua ambição e jovem energia, mas também intuição física, clareza de pensamento e um longo e apaixonado estudo dos raios cósmicos. (···) Antes da descoberta do decaimento píon-múon, Lattes e [Ugo] Camerini analisaram com muita precisão os produtos da desintegração do méson e mostraram que a energia total produzida era maior do que a equivalente da massa aceita para os então chamados mesotrons. Esse foi um trabalho de verdadeiro físico." [9].

Os ecos da versão de Occhialini em Schenberg e Fujimoto e a versão de Muirhead

Como veremos abaixo, a versão de Occhialini tornou-se a preferida entre os físicos brasileiros e entre aqueles que mantiveram de algum modo contato com o Brasil, como o físico japonês Yoichi Fujimoto. Contribuiu para isso o fato de Occhialini ter trabalhado entre 1938 e 1944 no Brasil – o que lhe permitiu tornar-se professor, colaborador e amigo de pessoas como Gleb Wataghin (1899-1986), Mario Schenberg (1914-1990), bem como do próprio Cesar Lattes – e também ter continuado a manter contato com esses personagens, mesmo após o seu retorno à Europa antes do fim da Segunda Guerra Mundial.

Na versão de Schenberg, lemos:

"Coube a Cesar Lattes o grande mérito de ter descoberto experimentalmente a existência dos mésons pi, confirmando as previsões teóricas do grande físico japonês Yukawa. Apesar do (*sic*) trabalho publicado ter sido firmado também por Occhialini e Powell [Hugh Muirhead também o assina], a ideia da utilização das emulsões nucleares para o estudo de raios cósmicos foi devida a Lattes, que teve também a ideia de levar as emulsões para expô-las nos Andes, a grande altitude, *como me informou Occhialini*. Powell se limitara a utilizar as emulsões nucleares para o estudo das reações nucleares." [10] (grifos nossos).

Também com base na versão de Occhialini, Fujimoto, um dos responsáveis pela Colaboração Brasil-Japão, vai além do fato de citar um único indivíduo, para atribuir o sucesso do trabalho em Bristol na área de raios cósmicos a uma escola que havia se formado no Brasil.

"Powell, the head of laboratory, knew that this venturing into the field of cosmic ray study was possible and successful only with participation of young school of São Paulo University." [11] (grifos nossos)

Como Powel e Occhialini, o físico Hugh Muirhead também esteve presente aos acontecimentos relativos à descoberta do méson pi no Laboratório H. H. Wills na Universidade de Bristol. Desse modo, suas palavras (ainda que escritas para um texto comemorativo dos 60 anos de Lattes) merecem atenção, porque nos fornecem detalhes acerca do que se passou naquela época. Com relação à participação de Lattes (e também de Ugo Camerini, outro físico brasileiro),

Muirhead nos conta o seguinte:

"The work in the group up to about Christmas 1946 mainly consisted of learning how to use this emulsion (Ilford C2 nuclear emusion) – calibrating it for range-energy purposes (mainly Giulio's responsability) and learning how to develop it." [12].

Gostaríamos de observar que no testemunho de Muirhead encontramos a afirmação de que Lattes também dava suporte teórico para o grupo de Bristol:

"The experience [análise das chapas que Occhialini expôs no final de 1946 no Pic du Midi] of our Brazilian friends [Lattes e Ugo Camerini] in cosmic ray work at this time was unvaluable and on many occasions I heard the name Wataghin mentioned ($\cdots$) *In putting this paper together [Nature, 24 de maio de 1947], we relied heavily on Giulio for help in basic theory as it existed at that time. I am sure the first time I heard the name Dirac mentioned was in the presence of either Giulio or Ugo.*" [13] (grifos nossos)

Como afirmamos anteriormente, não nos preocupamos em determinar que versão corresponde realmente aos fatos e nem mesmo explicar as contradições entre as narrativas que apresentamos acima. Mas pensamos ser importante assinalar a existência de versões que diferem bastante entre si. Enquanto em Powell e Lock a figura de Lattes é secundária (em Lock, o brasileiro é apresentado como antigo estudante de Occhialini) ou mesmo inexistente (autobiografia de Powell), nas outras versões percebe-se a presença do brasileiro como um interlocutor da mesma importância que os outros atores.

A versão de Cesar Lattes

Até aqui analisamos as versões de três dos quatro autores que assinaram o trabalho publicado na revista *Nature* de maio de 1947. Finalmente, falta-nos então apreciar as de Lattes. Nelas, é importante notar a ênfase dada ao desenvolvimento técnico (inclusão do bórax) das emulsões nucleares:

"*In the same experiment I placed borax-loaded plates, which Ilford had prepared at my request (c d o t s) Occhialini and I decided that he should take some plates* to the Pic du Midi in the Pyrenees for an exposure of about one month; some were loaded with borax, and some were normal plates (without borax)." [14] (grifos nossos).

"Bom, foi sorte (sic) a gente ter colocado bórax, porque, dentre todas as chapas que foram colocadas no alto do Pic du Midi, *as que não tinham bórax não registraram praticamente nada.* Havia um enfraquecimento da imagem latente e só se pegava o resultado dos últimos dias de exposição. *Mas aquelas com bórax mostraram, dentre outras, a partícula pi menos entrando na emulsão*, porque a ionização aumenta quando velocidade diminui." [15].

Não é nossa intenção afirmar que a versão de Lattes é a definitiva ou mesmo suficiente para que se possam compreender os reais papéis dos atores envolvidos.

Conclusão

"*Desse clima de trabalho, de amor à ciência, continuam a sair jovens pesquisadores e todos vós sabeis da importância e da repercussão mundial das pesquisas recentes de Cesar Lattes.*" (José Leite Lopes discurso pronunciado em 16/11/1948 ao tomar posse da cadeira de física teórica e superior da Faculdade Nacional de Filosofia da Universidade do Brasil).

Em uma palestra para um colóquio internacional sobre a história da física nuclear, realizado na década de 1970, o físico húngaro-americano Eugene Wigner (1902-1995) observou que contar a história é um processo seletivo, na medida em que personagens e eventos importantes em suas respectivas épocas, por vezes, acabam esquecidos anos depois (ver ref. [16]). A observação de Wigner é apenas mais uma maneira de se afirmar que são muitos os critérios de seleção nos processos narrativos da história (no caso, da ciência).

Nas várias narrativas que apresentamos sobre a detecção do méson pi, percebe-se a inclusão ou exclusão de certos elementos (nomes, locais, detalhes sobre a técnica, propriedade intelectual de ideias, contribuição dos participantes, entre outros) segundo o "local", a época e o enfoque adotados pelos diversos narradores. As necessidades e a própria relação desses narradores com a história determinariam a maior ou menor relevância de fatos e/ou personagens.

Não queremos afirmar que a história deva ser compreendida como sendo obrigatoriamente subjetiva. Deve-se, sim, reconhecer que a presença de elementos subjetivos na elaboração do processo de narrativa contribui para que a história possa vir a ser considerada como objetiva. Em outras palavras, a história não é feita sem um narrador.

No capítulo 5 de sua autobiografia (dedicado à detecção do méson pi), a versão de Powell organiza a narrativa de modo que a descrição das técnicas empregadas sobrepõe-se à nomeação e/ou contribuição dos personagens dessa história. A linguagem de Powell assemelha-se àquela de um relatório científico, em geral vista pelos próprios cientistas como o modo mais adequado (e "imparcial") de apresentação de teorias e fatos. Mesmo o físico italiano Giuseppe Occhialini, um cientista conhecido na época e principal colaborador de Powell, é poucas vezes mencionado. Ao favorecer as técnicas em detrimento dos personagens, Powell escreve uma história na qual a detecção do méson pi fica diluída na obra de um grupo (de físicos, químicos e microscopistas).

Para o Brasil daquela época, uma versão semelhante à apresentada pelo ganhador do prêmio Nobel de física de 1950 (na qual as conquistas são fruto de um grupo despersonalizado) com certeza não seria a mais adequada. Os físicos que aqui trabalhavam, brasileiros ou estrangeiros, optaram por contar a mesma história de modo diferente, mais "interessante" para os objetivos a que se propunham. As antigas e conhecidas deficiências de nossa estrutura universitária e as chances que agora surgiam com a detecção "exigiam" um outro tipo de narração. Para os interesses que norteavam um pequeno mas ativo grupo de cientistas da época era sem dúvida interessante que a participação de um brasileiro fosse ressaltada, não só para afirmar as qualidades científicas individuais do "nosso representante" em uma ciência que havia obtido prestígio no pós-guerra, mas também para fazer conhecer que este representante era "resultado" de esforços anteriores feitos por brasileiros e estrangeiros que aqui trabalharam.

Lattes é fruto do trabalho desenvolvido por Wataghin na Universidade de São Paulo desde 1934. Uma das principais preocupações do físico ítalo-russo foi a de sempre aliar o ensino à pesquisa. O laboratório do Departamento de Física da USP não era apenas um local onde experiências já conhecidas deveriam ser repetidas, como geralmente ocorre quando a finalidade do ensino é meramente didática. Mas deveria ser sim o ambiente no qual descobertas poderiam ser feitas. Seus alunos sempre foram "convidados" a praticar a pesquisa e a não se contentar com a situação de espectadores. Wataghin, consciente de que sua principal contribuição para o Brasil seria a formação de pesquisadores autônomos e bem-sucedidos (aptos a participar ativamente no processo de construção do conhecimento), transmitiu essa lição a seu alunos.

Os trabalhos de ensino e pesquisa desenvolvidos por Wataghin em São Paulo tornaram-se públicos (alcançando inclusive importantes personagens da esfera política da época) no momento em que o nome de seu antigo ex-aluno e colaborador Cesar Lattes começou a aparecer na imprensa.

A versão que destacava a participação do brasileiro Lattes num evento de repercussão mundial foi fundamental para convencer autoridades dos setores público e privado a investir na continuidade de uma escola que havia gerado cientistas e contribuições importantes não só para a ciência, mas também para o Brasil, na medida que o país, segundo o discurso desse pequeno grupo, teria a possibilidade de acompanhar o desenvolvimento científico e econômico no pós-guerra através daquele que era visto então como o principal domínio da ciência, a física nuclear.

Leite Lopes, Guido Beck (1903-1988) e o próprio Lattes souberam fazer uso dessa participação brasileira nos eventos de Bristol, Chacaltaya e Berkeley. Para eles, cientistas, não era suficiente a divulgação de seus nomes e feitos junto ao público em geral através da imprensa. O que também e mais lhes importava era extrair conquistas mais amplas, visando ao aperfeiçoamento de nossas instituições ligadas ao ensino e à pesquisa. Tal fato exigiu que eles mesmos narrassem intra e extramuros da academia a história da qual haviam participado, escrevendo artigos para jornais e revistas e pronunciando-se para seus pares – deste último caso são exemplos significativos o discurso de cátedra de Leite Lopes, em 1948 (citado na epígrafe desta conclusão), e o de Beck, no mesmo ano, para a Academia de Ciências de Buenos Aires.

Para que de fato obtivessem o sucesso que almejavam, Lattes, Leite Lopes e Joaquim Costa Ribeiro (1906-1960) perceberam a necessidade de entrar em contato direto com aqueles que poderiam tornar realidade suas reivindicações através do apoio político e financeiro. O que eles tinham em mente ao procurarem Álvaro Alberto, João Alberto Lins de Barros e o banqueiro Mário de Almeida, entre outros, era obter a introdução da dedicação exclusiva na estrutura universitária, bem como o aumento de recursos financeiros para a pesquisa e a continuidade na política de envio de jovens para o exterior a fim de completarem a formação recebida no Brasil, além da criação de cátedras e institutos de pesquisa.

Na verdade, em Lattes, o desejo de fortalecimento dessas estruturas já se mostra latente antes mesmo da detecção natural com as chapas do Pic du Midi e de Chacaltaya. Em carta a Leite Lopes, em 12/08/1946, Lattes escreveu:

"(···) Estou perfeitamente disposto a ir trabalhar aí em condições muito menos favoráveis do que aqui (estou me referindo à parte científica e à possibilidade material de pesquisa, não à parte profissional) porque acho que é muito mais interessante e difícil formar uma boa escola num ambiente precário do que ganhar o prêmio Nobel trabalhando no melhor laboratório de física do mundo".

Assim, coerentemente com a estratégia de aperfeiçoamento das condições para a realização da ciência no Brasil, os usos do feito de Lattes tinham que ressaltar que este último não era obra do acaso, mas sim representava a continuidade de um trabalho bem-sucedido, no qual muitos outros tomaram parte. E para isso foi preciso contar e difundir uma versão do descobrimento na qual a participação de atores brasileiros era importante. Na difusão dessa versão, contou com a colaboração de seus colegas, os quais, mesmo não tendo participado diretamente dos eventos, aqui se tornaram personagens tão importantes quanto ele, todos pertencentes a uma escola que deveria ser continuada. Um exemplo da difusão para o público geral no qual se enfatiza a participação do físico brasileiro está num trecho de um artigo escrito por Leite Lopes para comemorar os 70 anos de Lattes. Nele, lê-se:

"No Brasil, comecei a escrever sobre os trabalhos de Lattes; um primeiro trabalho saiu em 1947 [Lattes ainda estava na Inglaterra] no jornal Ciência para Todos, suplemento de divulgação científica de *A Manhã*, do Rio de Janeiro" [16] (itálicos no original).

Mais importante do que tentar saber qual a versão "verdadeira" e "definitiva" sobre a detecção natural e a confirmação em laboratório da existência do méson pi é perceber que os protagonistas brasileiros dessa história, para alcançar seus objetivos, souberam fazer uso da versão que ajudaram a elaborar e difundir. Nela, o principal papel de Lattes foi servir de elemento de ligação – e, principalmente, de continuidade – entre a física existente no Brasil antes da Segunda Guerra Mundial e a que se pretendia construir nos anos seguintes.

Notas e Referências Bibliográficas

[1] Powell, Cecil Frank. *Fragments of an Autobiography* Bristol, University of Bristol, 1987, p. 20.

[2] Powell. *op. cit.* p. 20.

[3] Leite Lopes, José. "Cesar Lattes, o CBPF e a nova física no Brasil" in *Cesar Lattes 70 anos – A Nova Física Brasileira*, A. Marques (Ed.), Rio de Janeiro, CBPF, 1994, p. 80.

[4] Lock, Owen. "Cecil Powell: pions, peace and politics", *Physics World*, november 1997, pp. 35-40.

[5] Lock, Owen. "Half a century ago – The Pion Pioneers", *Cern Courier*, vol. 37, no. 5, junho de 1997, pp. 2-6. Reproduzido neste livro, pp. 13-16.

[6] Powell, Cecil Frank; Fowler, Peter; Perkins, Donald. *The Study of Elementary Particles by the Photographic Method – An Account of The Principal Techniques and discoveries illustrated by An Atlas of Photomicrographics*, Pergamon Press, London, 1959, p. 26.

[7] Powell, Fowler e Perkins, *op. cit.* p. 32.

[8] *50 Anos da Descoberta do Méson pi: 1997/1998.* Catálogo comemorativo da exposição, São Paulo, 1998.

[9] Occhialini, Giuseppe. "Cesar Lattes, os anos de Bristol", *Ciência e Cultura*, vol. 36, no. 11, 1984, p. 2066-68.

[10] Schenberg, Mario, "Cesar Lattes, grande físico e personalidade extraordinária" in *Topics on Cosmic Rays – 60th anniversary of CMG Lattes*, vol. 1, Campinas, Editora da Unicamp, 1984, p. 11.

[11] Fujimoto, Yoichi. "Discovery of Pi and Mi Mesons and Brazil-Japan Collaboration on Cosmic Rays", *Ciência e Sociedade*, CBPF-CS-025/97, setembro de 1997, p. 3. Reproduzido neste livro, pp. 25-32.

[12] Muirhead, Hugh, "Encouters with Giulio Lattes" in *Topics on Cosmic Rays – 60th anniversary of CMG Lattes*, vol. 1, Campinas, Editora da Unicamp, 1884, p. 14.

[13] Muirhead, Hugh. *op. cit.*, p. 15. Essa situação é parcialmente corroborada com a seguinte declaração de G.D. Rochester: "*It is interesting to note the lack of theoretical support at Manchester and Bristol during the most fertile periods of discovery* (···) *At Bristol, too, although there were outstandingly good theoretical physicists in the University, there is little evidence that they made much impact on the Powell group, indeed, one always got the impression that the group was quite self-contained.*" (Rochester, G. D. intervenção oral, Colloque C8, Supplément au no 12, tome 43, décembre 1982, *Journal de Physique*, p. 189.)

[14] Lattes, Cesar, "My work in meson physics with nuclear emulsions" in *The Birth of Particle Physics*, Laurie M. Brown e Lillian Hoddeson (eds.), New York, Cambridge University Press, 1983, pp. 307-10, p. 308.

[15] Lattes, Cesar, "O nascimento das partículas elementares", Ciência e Sociedade, CBPF-CS-023-97, agosto de 1997, p. 11. Texto reproduzido a partir de vídeo com a palestra. A linguagem informal foi mantida na transcrição do vídeo e reproduzida no texto publicado. N.E.: Capítulo 12 deste volume.

[16] Citado por Olivier Darrigol em "Patterns of Oblivion: The Case of Guido Beck's Early Works" in *Guido Beck Symposium*, H. Moysés Nussenzveig e Antonio Augusto Passos Videira (eds.), *Anais da Academia Brasileira de Ciências*, vol. 67, suplemento 1, 1995, pp. 37-47, p. 37.

[17] Leite Lopes, José. *Op. cit.*, p. 81.

15

O Nascimento das Partículas Elementares

Cesar Lattes[1]

Recebi um telefonema do Presidente do CNPq, Prof. Tundisi, há uma semana, ou mais, me convidando para falar nesta cerimônia e, finalmente, anteontem recebi o convite oficial propondo o título "O Nascimento das Partículas Elementares". Este é o título de um Congresso que ocorreu em Batavia, no FERMILAB, nos anos 1980, envolvendo historiadores, físicos e outras pessoas, com a duração de uma semana.[2] Quer dizer, fica difícil em uma hora fazer a mesma coisa, não é? Vou, portanto, selecionar algumas coisas.

Eu gosto da Bíblia e lá no *Ecclesiastes* (7, 16) está escrito pelo Rei Salomão o seguinte: – "*Não busque ser demasiado sábio, nem demasiado justo. Você quer se arruinar?*". De maneira que eu não vou buscar esse caminho. Quanto ao título, acho que ele é um pouco estranho: "Nascimento" tem um significado antropomórfico, a não ser que sejamos animistas (eu sou um pouco animista); "Elementares", já sabemos que não são mais elementares; e o conceito de

---

[1] *N.E.*: Palestra de abertura da Comemoração dos 45 Anos do CNPq, realizada no IMPA, Rio de Janeiro, no dia 11 de outubro de 1996. Texto editado por F. Caruso e A. Troper, a partir da transcrição de vídeo feita por Sônia Ribeiro, e revisto pelo Prof. Lattes. O título da Conferência foi mantido, embora, como o leitor logo verá, o Prof. Lattes talvez tivesse optado por outro. Os editores gostariam de esclarecer que eles se renderam ao bom humor e ao estilo coloquial do seminário do Prof. Lattes e, portanto, na medida do possível, procuraram preservar estas características no texto escrito. Eles gostariam de agradecer ao Prof. Alfredo Marques pela colaboração na edição deste texto e a Roberto Alves pela digitalização das imagens.

[2] N.E.: *Cf.* Laurie M. Brown, Max Dresden and Lillian Hoddeson (Eds.), *Pions to Quarks*, Cambridge, Cambridge University Press, 1989.

"Partículas" é relativo...

Costuma-se iniciar a conversa sobre as partículas elementares com os Gregos, com Leucipo e Demócrito, cerca de 500 a.C., fundadores da Teoria Atômica, mas sem uma base empírica. Além desta corrente, havia outra, cujo maior expoente foi Aristóteles. A corrente atomista de Demócrito passou para o Império Romano através de Epicuro, e Lucrécio, o poeta, escreveu o *De Rerum Natura*, que deu origem à Teoria Atômica. Mas não foi só na Grécia que se discutiu o atomismo. Também na Índia e na China existiram pensadores que discutiram teorias atômicas.

Maimônides,[3] um judeu árabe, português, que viveu por volta do ano 1000, era atomista, mas não só para matéria; era também atomista no que se refere ao espaço e ao tempo. Eu gosto dessa ideia. Também Copérnico e Newton eram atomistas. Do início do segundo milênio até Copérnico e Newton a Ciência ficou meio parada, por causa da autoridade de Aristóteles.

Newton não se compromete ao dizer que "não fazia hipóteses". Ele atribui a Deus a criação dos átomos e diz que Ele fez os átomos da maneira que dessem para fazer o Mundo como ele queria.[4] Gassendi, mais ou menos contemporâneo de Newton, foi o primeiro que, em 1650, distinguiu átomo de molécula. Daniel Bernoulli deu o primeiro passo importante: deduziu que a pressão era devida à quantidade de movimento das moléculas batendo nas paredes do recipiente. Se continuarmos neste caminho chegaremos a Lavoisier. Antes dele, havia uma confusão muito grande entre misturas e compostos *etc.* Lavoisier definiu, de uma maneira operacional, que o que não podia se decompor era elemento e classificou uns 30. Além disso, descobriu, junto com Lomonosov, a conservação da massa, que é importante para se entender o resto da Física.

Um passo importante na Teoria Atômica, ou das Partículas Elementares, foi dado pela Química. Com os 30 elementos de Lavoisier, Proust enunciou a "lei das proporções definidas", *i.e.*, que as massas dos compostos estavam sempre num quociente determinado.

Dalton e Gay-Lussac mostraram que a "lei da composição dos gases" também envolvem números inteiros. A coisa foi posta em ordem por Avogadro com a hipótese de que todo gás, à pressão e à temperatura constantes, tem o mesmo número de moléculas. Foi o químico italiano Cannizzaro quem primeiro viu

---

[3] N.E.: Maimônides nasceu em Córdoba, Espanha, e viveu de 1135 a 1204.

[4] N.E.: "*All these things being consider'd, it seems probable to me that God in the beginning form'd Matter in solid, massy, hard, impenetrable, moveable Particles, of such Sizes and Figures, and with such other properties, and in such Proportion of Space, as most conduced to the End for wich he formed them; and that these primitive Particles being Solids, are incomparably harder than any porous bodies compounded of them; even so very hard, as never to wear or break in pieces; no ordinary Power being able to divide what God himself made one in the first creation.*" (Newton: *Opticks*).

claramente a distinção entre peso molecular e peso atômico e ampliou e aplicou a hipótese de Avogadro à Teoria Atômica.

A primeira ideia de átomo com um núcleo e alguma coisa ao redor é de 1828, mas a atração, era suposta gravitacional – foi Ampère que deu a ideia de forças elétricas. Em 1833, Faraday apareceu e, com a lei da eletrólise, ele chegou aos íons positivos e negativos. Stoney, um físico inglês, chegou a avaliar a carga quântica. Ele mostrou que para entender as leis da eletrólise era preciso supor que a carga era múltipla de pedacinhos sempre iguais e avaliou da ordem de 10^{-10} unidades eletrostáticas, talvez 10 vezes menos do que se admite hoje. Então, *elétron*, inicialmente, não denotava a partícula elétron; era o *quantum* de carga elétrica (1891).

Por outro lado, se tinha a Tabela Periódica de Mendeleiev para os elementos químicos. Mendeleiev e outros mostraram que havia uma periodicidade nos elementos e, inclusive, nas vagas dos períodos e previram a existência de outros elementos que foram descobertos mais tarde.

O átomo, de repente, deixou de ser uma coisa única, porque J.J. Thomson mediu a razão entre a carga e a massa do elétron (e/m) e mostrou que se tratava de uma partícula de massa aproximadamente 2 000 menor que a massa do hidrogênio. Thomson mediu e/m e Kaufmann mostrou que esta razão varia com a velocidade (diminui à medida que a velocidade aumenta). Isto ainda no século XIX.

Em 1900, o pessoal achava que a Física estava tão adiantada, como está agora, que era só uma questão de mais algumas casas decimais nas medidas; mas, de repente, Beckerel descobriu a radioatividade. Ele, Madame Curie e Pierre Curie mostraram que havia emissão de partículas de grande energia por certos elementos químicos e Mme. Curie mostrou que os chamados raios β eram elétrons negativos, enquanto Rutherford mostrou que as chamadas partículas α eram núcleos de hélio.

Por outro lado, Soddy mostrou, a partir das cadeias radioativas, que havia elemento de mesma propriedade química, mesmo número atômico, mas de massas diferentes. Foi ele quem introduziu o conceito de isótopo. Mais tarde, Astron conseguiu separar experimentalmente os isótopos. Utilizando campos magnéticos e elétricos ele separava vários valores de q/m e mostrou que certos átomos eram compostos de vários tipos de núcleos de massas diferentes, embora quimicamente quase idênticos (mesma carga nuclear).

Bom, o átomo era indivisível, mas de 1908 a 1911 Geiger e Marsden, que trabalhavam no laboratório de Rutherford, fizeram o espalhamento de partículas α de altas energia – emitidas por uma amostra de polonio – por vários elementos, em particular o berílio, e viram que, contrariamente ao modelo da moda (o modelo de Thomson), segundo o qual a carga positiva estava distribuída por todo ou quase todo o átomo de raio 10^{-8} cm, havia um caroço com quase toda a massa atômica, e raio da ordem de cem mil vezes menor que o do átomo. Esta pequena região é o *núcleo*, como foi interpretado por Rutherford. Assim, nasceu o átomo nuclear. Este átomo nuclear, pelas leis da eletrodinâmica, não seria estável e nisso Niels Bohr deu um jeito com seus postulados. Só a Mecânica Quântica mais tarde, em 1925-6, resolveu os paradoxos introduzidos por Bohr.

Rutherford também fez a primeira reação nuclear. Ele arrebentou não só o átomo; ele arrebentou o núcleo, mandando partículas α em nitrogênio obtendo oxigênio e prótons (núcleos de hidrogênio). Foi ele quem cunhou o nome *próton*, em 1920, na Reunião da Sociedade Inglesa de Física.

Nesse mesmo ano de 1920, Rutherford proferiu a *Bakerian Lecture* em uma importante reunião da Sociedade Inglesa de Física, cujo título foi: "O Nêutron". Sua ideia era a seguinte. Para entender a diferença de massa entre o oxigênio comum, que tem massa 16, e o outro de massa 17, obtido através do bombardeamento do nitrogênio com partículas α, ele imaginou que deveria existir uma partícula neutra de massa um pouco maior que a do próton. O nêutron só foi descoberto 12 anos mais tarde por Chadwick.

A emissão de raios β pelos núcleos trouxe problemas, pois o espectro de emissão β era contínuo; havia algumas linhas, mas o espectro de energia era contínuo e Niels Bohr chegou a cogitar que a lei da conservação de energia fosse violada em alguns processos elementares. Guido Beck, que posteriormente trabalhou no Brasil, invocando esta ideia de Bohr, elaborou uma teoria que descrevia razoavelmente o espectro do decaimento β. Mas Pauli não gostou disso e postulou a existência de uma partícula neutra, muito penetrante, que se encarregava de roubar energia. A esta partícula Fermi chamou de *neutrino* e usou esta ideia original na teoria dele de emissão β, em 1934, que funcionou bem.

O neutrino, na verdade, só foi detectado em 1953, porque ele é muito penetrante; ele atravessa facilmente o diâmetro da Terra. Então, foi preciso um reator de grande potência produzindo muitos neutrinos por decaimento β para provocar uma reação do tipo $v + p \rightarrow n + e^{+}$.

Sobre o elétron positivo – o *pósitron* – quero dizer que, em 1932, Anderson obteve em uma câmara de Wilson a fotografia mostrada na Figura 15.1. Como os físicos aqui são poucos, eu vou explicar a ideia geral desta câmara. Em um recipiente com um êmbolo que pode expandir um gás – que pode ser ar e um pouco de vapor d'água – se provoca a expansão rápida, à temperatura baixa, e a água tende a se condensar em gotas. A expansão tem que estar calibrada pois, se for muito pequena, o vapor não condensa, e se for muito grande dá uma nuvem. Com a expansão certa (que condensa em gotas), o rastro de ionização deixado pela passagem da partícula carregada na câmara condensa-se em gotas, e assim, se vê a trajetória da partícula. A fotografia está certa; neste caso, a partícula carregada vem de baixo atravessa uma camada sólida muito fina de chumbo e continua. Vocês podem ver que a curvatura é menor na parte inferior da foto da Figura 15.1 e é maior na parte superior, por isso se diz que ela vem de baixo.

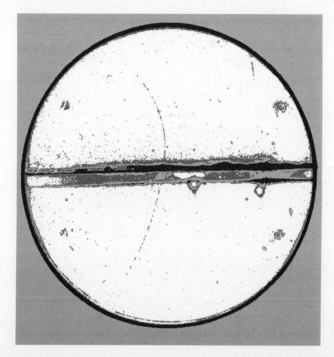

Figura 15.1: Fotografia (em negativo) do traço de ionização deixado por um pósitron em uma câmera de nuvem no experimento de Anderson, ao atravessar uma placa de chumbo de 6 mm. Extraída do livro de John Darius; *Beyond Vision*, Oxford Univ. Press; Oxford, 1984, que utilizou como fonte a foto do *Science Museum* de Londres.

Anderson já tinha um caso deste tipo (antes de colocar o chumbo no diâmetro da câmara) mas o Millikan, que era o chefe dele, dizia que eram elétrons que vinham de baixo. Quando Anderson apresentou esta fotografia em Cambridge, o Rutherford perguntou se ele tinha certeza que não tinha mudado o campo magnético. Este é o *pósitron*, que dizem que Dirac havia previsto na teoria dele, mas, na realidade, quando ele obteve a sua equação linear, relativística, ele obteve soluções de energia negativa e positiva, mas ele achava que as soluções de energia negativa correspondiam aos prótons. Foi preciso gente como Oppenheimer explicar para ele que não dava certo; se aquelas soluções fossem prótons, o hidrogênio seria instável; teriam que ser partículas da mesma massa que o elétron mas carga positiva, como os pósitrons.

Então o nêutron e o pósitron, que são mais ou menos descobertos simultaneamente, são as duas partículas elementares que não nasceram na década de 1930, estavam lá.

É interessante que no caso do nêutron a detecção foi da seguinte maneira: bombardeando o Berílio com partículas α, Bothe e companheiros, obtiveram uma radiação muito penetrante e Irène e Joliot Curie pensaram que fosse raio X, que dava Efeito Compton nos prótons; só que este efeito nos prótons é muito pequeno, não dava. Chadwick, que conhecia a aula do Rutherford de 12 anos antes, fez imediatamente a experiência correta e mostrou que tratava-se de uma partícula de massa um pouquinho maior que a do próton. Isso foi publicado e o grupo de Berkeley que dispunha de um Ciclotron de 1 milhão de volts e de deutério, que Urey havia separado, pode bombardear com deutério vários alvos, e obtinha sempre um grupo de prótons. Os físicos deste grupo achavam que era o nêutron que se soltava e obtiveram a massa diferente da do Chadwick. O Rutherford pediu uma amostrazinha de água pesada e bombardeou um alvo de água pesada com água pesada e mostrou que não eram o que os americanos diziam. Não é que a massa do Chadwick estivessse errada; eles estavam detectando, na verdade, o resultado de 2 reações: *deuteron* + *deuteron* $\rightarrow$ H_1^3 + *próton* e *deuteron* + *deuteron* $\rightarrow$ He_2^3 + *nêutron*, que, hoje em dia, são reações bem conhecidas porque são importantes para a bomba de Hidrogênio.

Logo depois que o Anderson detectou o pósitron, Occhialini foi trabalhar em Florença. Aprendeu a fazer contadores Geiger, que contam partículas elementares ionizantes, e a fazer coincidências eletrônicas com Rossi. Depois ele foi para Cambridge para trabalhar com Blackett, que tinha uma câmara de Wilson há muito tempo, e montou um dispositivo para disparar e fotografar automaticamente o que a câmara de Wilson mostrava. Quando a câmara de Wilson expande, formam-se gotinhas aonde tem carga e foi assim que se descobriram os primeiros chuveiros com elétrons e pósitrons (*Cf.* Figura 15.2), que são os chuveiros eletromagnéticos.[5] Bom, essas eram as partículas elementares até 1934.

Em 1934, Yukawa, sabendo que o núcleo, conforme medido por Rutherford, tem uma massa igual a quase toda a massa do átomo, e o raio da ordem de 10^{-13} cm, tentou entender as forças que mantinham o próton e o nêutron ligados, porque 2 prótons de cargas positivas deviam se repelir, enquanto dois nêutrons não se atraem. Ele postulou que, assim como o fóton é intermediário da interação eletromagnética, havia alguma coisa que era intermediária da interação nuclear: é o *méson de Yukawa*, e calculou que a massa deste méson deveria ser cerca de 200 massas eletrônicas; isso foi, repito, em 1934.

Em 1937, Anderson & Neddermeyer e Street & Stevenson (dois grupos independentes) obtiveram fotografias que mostravam a existência de partículas com massa de mais ou menos 200 vezes a massa do elétron. Os teóricos, que não tinham dado

---

[5] N.E.: Blackett e Ochialini deram, pouco mais tarde, uma importante contribuição ao estudo e à interpretação destes tipos de "chuveiro", que resultam da materialização de raios gama (fótons de alta energia) de origem cósmica em um par elétron-pósitron, fenômeno este que só pode ocorrer, de modo a satisfazer a conservação de momento e energia, na vizinhança de núcleos.

bola até então, de repente se entusiasmaram e fizeram as *teorias mesônicas* das forças nucleares. O Anderson chamou essa partícula de *mésotron*, porque ele não

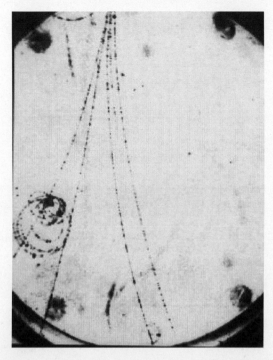

Figura 15.2: Negativo da fotografia feita por Anderson no topo de uma montanha no Colorado, mostrando a criação de um "chuveiro" de 3 elétrons e 3 pósitrons a partir de raios cósmicos. Os elétrons se curvaram para a esquerda e os pósitrons, para a direita. Extraída do livro de Frank Close, Michael Marten & Christine Sutton, *The Particle Explosion*, New York, Oxford University Press, 1987, p. 74.

sabia grego; mas devia ser *méson*, intermediário (quanto à massa) entre próton e elétron. Na verdade, Anderson acabou acertando porque se viu depois que o méson μ (como é conhecida hoje esta partícula) não é um méson, mas sim um "elétron pesado".

O próprio Yukawa previa que o méson era instável e desintegrava-se em elétron e mais alguma coisa.

Mas, como estamos no Brasil, vamos falar um pouco do que foi feito no Brasil. A Física Moderna no Brasil iniciou-se em 1934, com a Fundação da USP e com o contrato de especialistas estrangeiros. Teodoro Ramos, um bom matemático, que foi o encarregado de organizar a Universidade, achou que devia ir buscar na Europa os professores. Para a Física consultou Enrico Fermi que indicou o professor Gleb Wataghin (Figura 15.3, que pode ser considerado o pai da Física Moderna brasileira.

Figura 15.3: Prof. Gleb Wataghin. Foto gentilmente cedida pelo CBPF.

Russo de origem, formado na Ucrânia, durante a guerra civil, por ser filho de nobres, foi parar em Turim e tocava piano no cinema até que o Professor Perucca ofereceu a ele um cargo na Escola Politécnica e depois no Instituto de Física.

O Prof. Wataghin sabia que há dois comprimentos básicos na Física. Um deles relaciona-se ao raio clássico do elétron, ao comprimento de Compton do μ e o alcance da força nuclear (ou o raio nuclear) que são todas grandezas da ordem de 1 F:

$$\frac{e^2}{m_e c^2} \approx \frac{h}{m_\mu c} \approx \frac{q^2}{m_N c^2} \approx 10^{-13} \text{ cm} = 1 \text{ F}$$

onde e é a carga elétrica do elétron ($e^2/\hbar c \approx 1/137$), q é a carga da interação nuclear ($q^2/\hbar c \approx 15$ era o valor da época), h é a constante de Planck e, m_e, m_μ e m_N são, respectivamente, as massas do elétron, do múon e do núcleo. Ele cismou que não era coincidência, que isto era um comprimento fundamental. Logo ele chegou ao Brasil e publicou no *Zeitschrift für Physik* um trabalho aonde, a partir da ideia que acima de certa energia era preciso tomar cuidado com a Física, ele previu a produção múltipla de partículas: múltipla, não plural, *i.e.*, numa colisão só, várias partículas são produzidas. Isso foi em 1934, antes, portanto, do méson de Yukawa, que foi de 1935. E quando saiu o méson de Yukawa, Wataghin disse: "é produção múltipla de méson."

Aluno do Prof. Wataghin, Marcello Damy de Souza Santos (*Cf.* Figura 15.4), uma vez formado, foi à Inglaterra, aprender a fazer contadores rápidos de partículas elementares.

Figura 15.4: Prof. Marcello Damy de Souza Santos. Foto gentilmente cedida pelo Prof. Alfredo Marques.

O Prof. Marcello, inclusive, inventou um circuito muito mais rápido, dez vezes mais rápido do que se usava na Inglaterra, e então foi feita esta experiência aí (Figura 15.5).

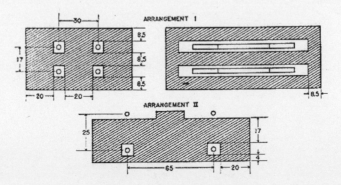

Figura 15.5: Arranjo do telescópio no experimento realizado a 800 m de altitude. Os blocos absorventes são de chumbo. Extraído de *Phys. Rev.* 57, p. 61 (1940).

Quatro contadores em coincidência, exigindo que os quatro fossem disparados ao mesmo tempo, com um poder de resolução de 10−7 s, enquanto o aparato inglês permitia uma resolução de 10^{-6} s. Podia-se mostrar que coincidências casuais, *i.e.*, coisas independentes que atravessassem os quatro contadores eram desprezíveis. Wataghin conclui, então, que era produção múltipla de mésons. Múltipla quer dizer: numa só colisão saem vários mésons, enquanto que plural é uma cascatinha, um méson de cada colisão. Isso foi confirmado uns 6 meses depois por um grupo inglês, liderado pelo físico húngaro Janossy, só que eles achavam que era produção plural. Essa controvérsia continuou por 10 anos e só quando apareceram máquinas como o Cósmotron, que acelera partículas acima de 1 bilhão de volts, é que viram que era produção múltipla mesmo. Mas Wataghin, mesmo antes disso, junto com Sala, fez experiências colocando água em cima dos detectores e mostraram que havia produção múltipla de mésons em hidrogênio, e com hidrogênio não dá para ser plural; quer dizer, se fosse Carbono tinha uma cascatinha, mas hidrogênio tem que ser múltipla. Logo, a produção múltipla foi descoberta no Brasil, em 1940.

Bom, se o pai da Física brasileira foi Wataghin, a mãe foi Giuseppe Occhialini, ou vice-versa (*risos na plateia*). Occhialini era antifascista e depois que terminou o trabalho dele com Blackett, na Inglaterra, voltou para a Itália, mas estava quente e, numas férias do Wataghin em Turim, o pai do Occhialini, que era Diretor do Instituto de Física de Genova, pediu ao Wataghin para dar um jeito de levar seu filho para o Brasil, senão ele iria se meter em encrenca. E Occhialini veio para o Brasil; é esse Senhor que se vê na fotografia a seguir (Figura 15.6), falecido em 1993.

Figura 15.6: Prof. Giuseppe Occhialini. Foto gentilmente cedida pelo CBPF.

Occhialini foi para a Inglaterra ainda durante a Guerra como voluntário, para combater o pessoal do eixo, porque era antifascista. Chegou lá e o colocaram para lavar pratos no restaurante militar. Foi preciso um telefonema para o Blackett, a partir do qual mandaram-no para a Universidade de Bristol – uma universidade particular pequena –, aonde estava Powell, que era um pouco esnobado porque tinha se recusado a fazer serviço de guerra e era considerado bem esquerdista. Powell trabalhava com emulsões fotográficas, mas ele era conservador, e, naquele tempo, eram ainda as emulsões fotográficas que se usavam para fotografia do dia a dia e tinham sido usadas em Física Nuclear por Marietta Blau e por Kinoshita, no Japão, no começo do século.

Na Figura 15.7, a mais à esquerda é a emulsão fotográfica que Occhialini viu o Powell usar. É até uma fotografia fácil por se tratar de um traço praticamente rasante à superfície bem plana da emulsão, de maneira que está toda em foco, mas, quando a trajetória da partícula está inclinada em relação ao plano da emulsão é duro de ver. Mas Occhialini tinha muita imaginação e tinha medo de estar vendo prótons aonde não havia. Ele então telefonou para a fábrica *Ilford* e saiu

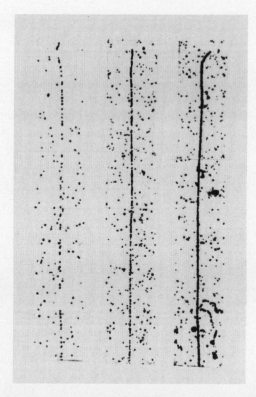

Figura 15.7: Traços deixados por prótons em diversas emulsões na proximidade do final de seus alcances. Extraído do livro C.F. Powell, P.H. Fowler and D.H. Perkins, *The Study of Elementary Particles by the Photographic Method*, London, Pergamon Press, 1959.

essa outra fotografia que está melhorzinha (foto do centro na Figura 15.7): é a chamada emulsão C2, de 1946. Esse detector é 2 000 vezes mais denso que a câmara de Wilson e, portanto, pode parar muito mais coisa por unidade de volume. Em compensação, com o campo magnético muito intenso, os traços são muito curtos. Mais tarde, fizeram emulsões sensíveis ao mínimo de ionização (foto mais à direita na Figura 15.7).

Na época, em que Occhialini tinha estado no Brasil, ele trouxe uma câmara de Wilson, igual à do Blackett, mas não se conseguiu colocá-la para funcionar. Eu tinha resolvido trabalhar em Física Experimental, porque com o Wataghin e Schönberg era muita conta; só a lagrangiana de um campo eletromagnético de um elétron puntiforme com momento de dipolo tinha 99 termos! Esta minha decisão levou-me ao Occhialini e lhe disse que gostaria de trabalhar com ele. Occhialini, durante a guerra, havia sido guia de montanha em Itatiaia, porque era considerado inimigo, sendo italiano, mas depois ele precisou de dinheiro para viajar para a Inglaterra e veio a São Paulo dar um curso sobre Raio X. Quando eu falei que queria trabalhar com ele, ele me disse para esperar um pouquinho. Foi lá dentro do laboratório e depois de um tempo voltou com um tubo de ensaio na mão e me disse para segurá-lo. Eu o segurei e queimei o meu dedo. Perguntei a ele qual era a ideia. Occhialini respondeu: – "É para te ensinar a não confiar em ninguém." (*risos na plateia*).

Occhialini foi então para a Inglaterra e eu consegui colocar uma outra câmara para funcionar, porque a dele estava envenenada. Mandei-lhe uma fotografia, ele mandou de volta fotografias como essas que vocês viram anteriormente (Figura 15.7) e eu vi que se eu estava querendo ver mésons parando, a emulsão era duas mil vezes mais eficiente e pedi para ele me arrumar um lugar lá em Bristol. Ele conseguiu um auxílio de 15 libras por mês, dado pela fábrica de cigarros Wills – foi aí que eu comecei a fumar –, e a Fundação Getulio Vargas, com o prestígio do Leopoldo Nachbin, me pagou a viagem num cargueiro.

Isto foi logo depois da Guerra e o cargueiro levou quarenta dias para chegar em Liverpool. Lá constatei uma coisa estranha: o Powell era conservador e continuava a usar as chapas antigas e essas chapas novas ficavam em cima da mesa.

Então me disseram para determinar a relação entre o alcance, o comprimento do traço, e a energia. Eu fiz isso com as reações nucleares de dêuterons com níquel, berílio e boro no gerador Cockcroft-Walton de Cambridge, de um milhão de volts, mas também carreguei chapas com bórax[6] (Figura 15.8).

Então eu tive a ideia de expor as emulsões em montanha, para determinar a energia de nêutrons cósmicos. Como Occhialini ia para o Pic du Midi, a 2 800 metros de altitude, pedi a ele que levasse algumas emulsões. O traço mais a

---

[6] N.E.: Tetraborato de sódio.

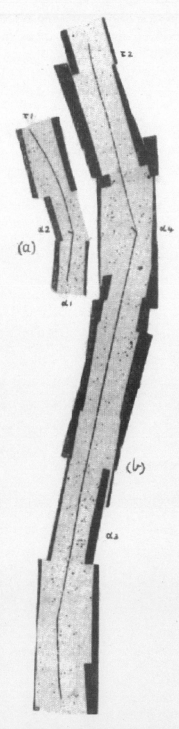

Figura 15.8: Mosaicos de fotomicrografias que permitiram a determinação da energia e momentum de nêutrons rápidos em Raios Cósmicos. *Nature*, March 8, (1947).

esquerda (Figura 15.8.a) corresponde ao evento obtido em laboratório, de 13 milhões de volts, e o da direita (Figura 15.8.b) ao evento de 40 milhões de volts, este último obtido na montanha. A partir dele dá para determinar a energia e a quantidade de movimento da partícula incidente e constatar que ela tinha a massa de nêutron.

Bom, foi sorte a gente ter colocado bórax, porque dentre todas as chapas que foram colocadas no alto do Pic du Midi, as que não tinham bórax não registraram praticamente nada. Havia um enfraquecimento da imagem latente e só se pegava o resultado dos últimos dias de exposição. Mas aquelas com bórax mostraram, dentre outras, a partícula π^- entrando na emulsão porque a ionização aumenta quando a velocidade diminui. Sabe-se que ela é mais leve do que o próton pela curvatura e deve ser negativa (Figura 15.9) porque ela entra no núcleo e arrebenta, transformando massa em energia, liberando cerca de 250 milhões de volts. Portanto, foi graças ao bórax que passamos a perna nos outro laboratórios.

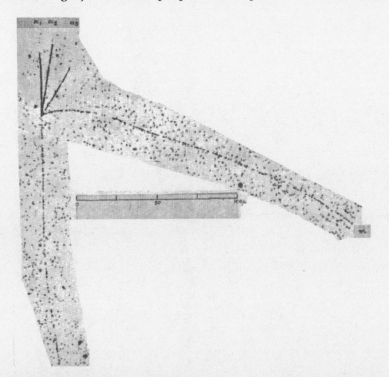

Figura 15.9: Méson π^- capturado por C_6^{12}, *Nature* (May 1946), p. 697.

Agora, uma moça do nosso grupo, chamada Marietta Kurz, depois de um tempo, descobriu o evento mostrado na Figura 15.10.

A trajetória é "tortinha" como deve ser a do méson que tem massa menor que o próton, sofre espalhamento múltiplo, para num certo ponto e sai outra partícula

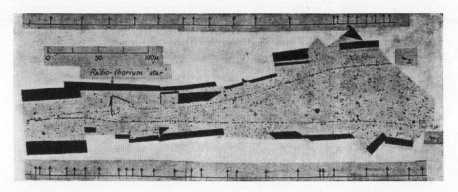

Figura 15.10: Desintegração espontânea de méson π^+ em μ^+. Nature, May 24, 1947, p. 695.

que, infelizmente, saía da superfície da emulsão, mas, pela quantidade de grãos, sabíamos que a trajetória estava quase no fim. Uma semana depois, foi vista a outra, pela Sra. Roberts, aonde estava tudo dentro do campo da emulsão e as duas tinham aproximadamente 600 mícrons de comprimento (Figura 15.11).

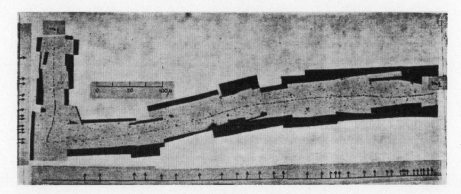

Figura 15.11: Como na Figura 10. Ibid.

Estava mais ou menos óbvio que se tratava de um processo fundamental e que havia dois tipos mésons. Nós não sabíamos que, em 1942, um dos colaboradores de Yukawa tinha verificado que a produção de mésotron era rápida (interação forte) mas a absorção debaixo da terra era lenta (interação fraca).

O Sakata viu logo que não dava para entender o fenômeno: se era produzido fortemente, deveria ser absorvido fortemente. Então ele supôs, e publicou em 1942, a teoria de dois mésons. O primeiro seria o méson de Yukawa que decaía no segundo, que era o penetrante. E é o que vimos aqui, mas na época nós não sabíamos.

Esses resultados foram obtidos com chapas expostas nos Pireneus, a 2 800 m de altitude. Como havia outros pesquisadores expondo chapas em avião, achamos que deveríamos andar rápido. Assim, fui ao Departamento de Geografia de Bristol e vi que a 20 km de La Paz, tinha um clube Andino e que se podia ir de automóvel a 5 500 metros de altura – *eighteen thousand feet* – (Chacaltaya). Então me pagaram uma viagem até lá e eu levei comigo as chapas até o topo onde foi feita a exposição. Quando tem neve o meio de transporte para chegar ao topo é um burrinho de cargas, que batizei de meu auxiliar.

Figura 15.12: Cesar Lattes no caminho para Chacaltaya.

Bem, as chapas ficaram um mês lá, depois eu fui buscá-las, revelamo-nas e apareceu muito mais coisas. Na foto reproduzida na Figura 15.13 vocês vêm um méson criado por raios cósmicos. Vocês estão vendo o "nascimento" do méson numa colisão com o núcleo, que se arrebenta e o méson de carga negativa para e transforma sua massa em energia e arrebenta outro núcleo.

Bom, vamos mostrar mais dois π e μ. Esses eventos são mais claros (Figuras 15.14 e 15.15).

As faixas pretas são um truque do Prof. Occhialini. Isso é um mosaico de fotografias porque estão em diferentes planos focais. Sem a faixa preta daria a impressão de tapeação, de que está tudo bem focalizado. Então, em várias

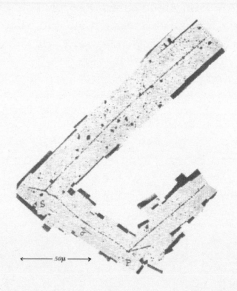

Figura 15.13: Emissão e absorção de méson π^-. *Nature*, October 11, 1947, p. 490.

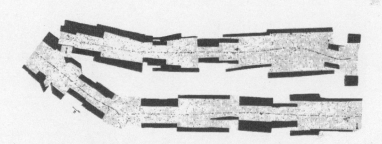

Figura 15.14: *Nature*, October 4, (1947).

Figura 15.15: *Ibid*.

fotografias ele deixou a faixa preta, para ser honesto, para mostrar que era um mosaico. Novamente o aspecto é o mesmo e, ao todo, foram detectadas umas 30 partículas. Contando os grãos, cuja densidade é proporcional à ionização, deu para mostrar que a massa do π que para é maior do que a do π que sai, da ordem de 1,4 para l; a massa que entra é da ordem de 300 e a que sai da ordem de 200 massas eletrônicas.

A última partícula descoberta, agora a gente lê no jornal – eu não leio mais revista científica, porque é demais –, ou melhor, a última partícula para a qual há indicação de que ela existe é chamada de quark top, com massa 140 milhões de volts, aproximadamente. O artigo é constituído de 4 páginas escritas e 2 páginas de assinaturas dos autores e nomes das instituições.[7] Eu quero mostrar, na Figura 15.16, que o grupo aqui era menorzinho (*risos na plateia*).

Figura 15.16: "O Grupo de Bristol". Fotografia gentilmente cedida pelo CBPF.

O chefe inglês, o Powell, está na extrema esquerda; a secretária dele que está em baixo era a própria mulher dele, que batia à máquina; as três moças são microscopistas. O careca é o *factotum*, o outro é fotógrafo, o barbicha é penetra (*risos na plateia*) e a da ponta é a Marietta Kurz, que viu o primeiro méson π. E os físicos são os de baixo: na frente, Muirhead e Goldschmit (que infelizmente morreu), Occhialini, Ritson, Locke, King, eu e Camerini e tem o mecânico também. Era um grupo relativamente pequeno e dividimos o trabalho. Nós – Lattes, Muirhead, Occhialini e Powell –, medimos a massa por contagem de grãos e os

---

[7] N.E.: Trata-se do artigo da descoberta do top pela Colaboração DZERO, da qual faz parte um grupo brasileiro do LAFEX/CBPF. Cf. Abachi, S. *et al.*, *Phys. Rev. Lett.* **74**, 2632 (1995).

outros 4 mediram pelo espalhamento coulombiano múltiplo: o méson se desvia de uma linha reta e dá para calcular a sua massa. Isso foi publicado em outubro de 1947. Em dezembro de 1947, Rochester e Butler, ex-alunos de Blackett, usando uma câmara de Wilson que era do Blackett, com controle automático, publicaram duas fotografias. Eu só tenho uma aqui.

A fotografia reproduzida na Figura 15.17 mostra esse V aqui na parte inferior direita da foto que, exatamente por isto, foi chamado partícula V. Isto é de dezembro de 1947 e o nosso trabalho é de outubro do mesmo ano. Eles concluíram que era alguma coisa neutra que decaia em um próton e em um méson negativo dos nossos; chamava-se partícula V.

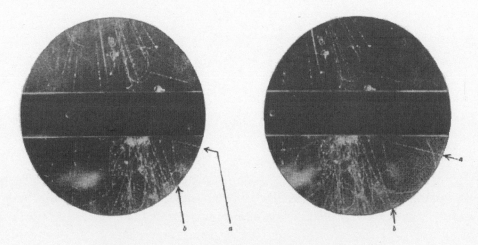

Figura 15.17: Produção das chamadas partículas V.

Outra fotografia revelou as partículas K, chamadas de partículas estranhas. Estas partículas eram chamadas de estranhas pelo seguinte: elas são produzidas fortemente e decaem lentamente, *i.e.*, para poder ser vista na câmara de Wilson tem que ter uma vida média maior que 10^{-10} s, e isso é lento, em Física Nuclear onde os tempos característicos são da ordem de 10^{-23} s. Abraham Pais, que foi citado hoje, levantou a hipótese de que devia haver produção associada que só podia produzir duas partículas estranhas de cada vez e haveria algum número quântico a ser conservado. Desta forma, na produção não haveria problema, mas no decaimento seria uma partícula estranha que ia decair em outras não estranhas, e isso não poderia acontecer rapidamente. Isto foi chamado de produção associada.

Bom, deixa eu comentar rapidamente como é que foram produzidos artificialmente 3 mésons na Califórnia. Lá havia um Ciclotron que podia acelerar partículas α até 400 milhões de volts (380 MeV, na verdade).

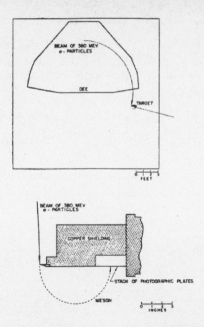

Figura 15.18: Esquema do aparato no Ciclotron de Berkeley. *Science*, March 12, (1948), vol. 107.

A sua construção foi iniciada logo após a descoberta do mésotron por Anderson, com doação da Fundação Rockefeller. Mas, durante a guerra, foi usado para separar urânio 235 e pulverizar em cima de Hiroshima. Terminada a guerra botaram-no para funcionar e, em novembro de 1946, ele estava com plena carga. No entanto, o pessoal lá acreditava em um outro méson com massa menor e não conseguiram detectar o méson π. Foi só depois que eu cheguei que os convenci que tinham que procurar um méson de massa maior e assim detectamos o méson π em laboratório. Deu certo.

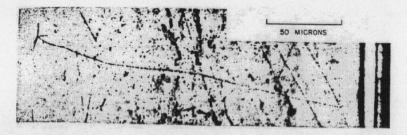

Figura 15.19: Méson π^- artificial. *Science*, March 12, (1948), vol. 107, p. 271.

Depois dos primeiros trabalhos foram feitas chapas aonde podia se ver mínimo de ionização. Assim ao invés de vermos só o méson pesado, decaindo no leve, vimos o pesado decaindo no leve e do leve sai o elétron.

Figura 15.20: Fotos de dois mésons produzidos artificialmente. *Ibid.*

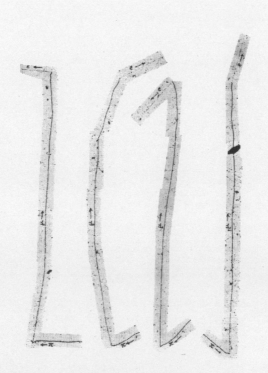

Figura 15.21: Decaimento sucessivo do méson π: $\pi \to \mu \to e$. Extraído do livro C.F. Powell, P.H. Fowler and D.H. Perkins, *Op. Cit.*, p. 245.

Para explicar porque a produção era forte e o decaimento era lento, Pais propôs que era produção associada. Tinha um novo número quântico, chamado estranheza, que devia se conservar. Então o méson π não é estranho mas dá o K^0 e o Σ^0 que têm estranheza oposta. Bom, nessa época o Sakata já tinha previsto que havia dois tipos de mésons e previu também que o μ não decaia em um elétron mais um neutrino (e ν). Isto só foi possível mostrar alguns anos depois.

O Prof. Leite Lopes, a par do que estava acontecendo, em 1958, publicou um trabalho fundamental, onde ele tentava unificar a interação eletromagnética, bem conhecida, com a interação fraca, que é responsável por essas desintegrações lentas e postulou, chegou à conclusão de que devia haver fótons intermediários mas de massa grande (os bósons intermediários). Ele se assustou porque sua estimativa deu 60 massas nucleônicas, mas publicou, sendo o primeiro. Mais tarde, Gell-Mann e Feynman fizeram essa teoria e finalmente esses bosóns forma encontrados, no CERN, em 1983, pelo grupo liderado pelo Rubbia, usando um acelerador de bilhões de volts.

Figura 15.22: Leites Lopes com Cesar Lattes, no CBPF em 1994, por ocasião das solenidades em homenagem aos setenta anos de Cesar Lattes.

Eu tentei mostrar para vocês que essa coisa não é só imaginação, não é só teoria. Para Demócrito era só imaginação, mas aqui tem coisas palpáveis. Não se sabe se antes do Prof. Lederman detectar o Υ, se o Υ existia; não se sabe, isso não se pode provar, mas é fé da gente, que há uma realidade objetiva. Berkeley não acreditava nisso.

Bom, as coisas foram se complicando. Apareceram, assim, além do raio γ, os bósons intermediários W^+, W^- e Z^0.

Sakata fez uma proposta de uma teoria aonde as partículas elementares tinham a estrutura composta. Ele não foi o primeiro, pois o primeiro foi de Broglie, que propôs que o fóton era um par de neutrino-antineutrino. Depois o Fermi e o Yang propuseram que o méson π era um par de nucleon e anti-nucleon. Quando eu perguntei ao Fermi se ele acreditava naquilo, ele disse: "*There is a fifty-fifty chance that it is right.*" (risos na plateia).

Mas Sakata pegou o que se conhecia na época, um ur-próton (p^*), o ur-nêutron (n^*) e o ur-lambda (Λ^*), que eram as partículas elementares conhecidas na época (Conferência de Kiev, de 1959), e supôs que o estado $p^*\bar{n}^*$ fosse o méson π, que $p^*\bar{\Lambda}^*$ desse para fazer o méson K e três deles dessem para fazer o nucleon e, deste modo, conseguiu uma classificação razoável das partículas conhecidas na época. Mas depois ele achou, usando Teoria de Grupos, que precisava acrescentar mais uma partícula, e previu a existência de outra, que foi descoberta pelo Niu, que a chamou de partícula X. Isso foi anterior à descoberta do charm, descoberto na América do Norte um pouco depois, com colisão elétron-pósitron, que dava uma ressonância e não era nem protônica, nem neutrônica, nem estranha: era charmosa, era diferente.

Bom, não vou falar dos vários modelos, falo do último, o último modelo é o modelo de Quarks, é o nome que o Gell-Mann deu, está ligado ao livro de J. Joyce.[8] Seriam partículas que têm carga elétrica igual a 1/3 ou a 2/3 da carga eletrônica, dizem que há 3 pares:

up	strange	top
down	charm	bottom

e são pesadas, e de acordo com os autores dessa teoria, dá para se construir uma descrição das partículas elementares, com exceção das leves (dos léptons): e, μ, τ, ν_e, ν_μ, ν_τ e dos bósons.

O pessoal acha que com isso dá pra descrever o Universo até 10^{-43} s depois da Criação, mas não dá muito certo porque há, pelo menos, 20 parâmetros que têm que ser obtidos da experiência. Além disto já são muitas partículas de novo e, além do mais, os quarks têm 3 cores (um tipo de carga relacionado à interação forte)

---

[8] N.E.: Quando este nome foi escolhido por Gell-Mann, em 1963, ele só tinha em mente o som da palavra, e não a grafia; algo como "kwork". Mais tarde ele encontrou a palavra quark no seguinte trecho do livro Finnegan's Wake de James Joyce:
"*Three quarks for Muster Mark!*
Sure he hasn't got much of a bark
And sure any he has it's all beside the bark."

cada. Lavoisier tinha só 30 partículas, né? Aqui são mais de 50!

Bom eu tentei dar uma ideia da evolução da Física de Partículas, dentro do limite de tempo que dispunha. Maiores detalhes poderão ser encontrados em um livrinho bom publicado pelo Santoro e pelo Caruso.[9]

Para concluir, gostaria de dizer que quando eu voltei da Califórnia com os mésons artificiais, não dava para fazer um Cíclotron no Brasil e, então, nós fomos para a Bolívia de novo e continuamos estudando raios cósmicos. Atualmente existe uma cooperação Brasil-Japão que estuda, há mais de 30 anos, interações de 10^{16} eV, 10^{16} eV e 10^{18} eV, – sempre 10 a 100 vezes acima das máquinas –, e se encontram coisas novas.[10] Não vou entrar em detalhes aqui e só vou mostrar o caso típico de produções múltiplas de mésons, prevista pelo Wataghin.

Essa partícula incidente, o núcleo arrebenta e há o cone para frente e para trás no centro de momentum e você tem no Lab., fica o cone fechado e aberto, produção múltipla, aquela do Wataghin, Souza Santos e Pompéia, de 1940. Só que nós agora estamos vendo energias bem maiores do que essas, o ângulo de abertura 10^{-5}, 10^{-6} rad. Então não se pode ver assim; é necessário colocar chapas de chumbo e olhar as cascatas eletromagnéticas dos raios γ nos mésons neutros.

Então isso é para dar uma ideia do que os físicos fazem atualmente para fazer jus ao salário de fim de mês. E, para terminar, como eu comecei com Salomão, vou citar Salomão. O *Ecclesiastes* (12, 12) diz assim:"*Não te metas a escrever livros nem estudes muito. Cansa a carne*".

Muito obrigado.

PUBLICAÇÕES DE CESAR LATTES NO PERÍODO 1947-1957[11]

1. C.M.G. Lattes, H. Muirhead, G.P.S. Occhialini and C.F. Powell, "Processes involving charged mesons", *Nature*, 1947; 159: pp. 694-697.

2. C.M.G. Lattes and G.P.S. Occhialini, "Determination of the energy and momentum of fast neutrons in cosmic rays", *Nature*, 1947; 159: pp. 331-332.

3. C.M.G. Lattes, G.P.S. Occhialini and C.F. Powell, "Observation on the tracks of slow mesons in photographic emulsion", *Nature*, 1947; 160: pp. 486-492.

---

[9] N.E.: Referência ao livro contendo as contribuições à sessão da LISHEP93 dedicada a professores de segundo grau e licenciandos: F. Caruso & A. Santoro (Eds.), *Do Átomo Grego à Física das Interações Fundamentais*, Rio de Janeiro: AIAFEX, 1994.

[10] N.E.: Podemos citar como exemplo os eventos Centauro.

[11] N.E. Achamos oportuno reproduzir aqui as referências dos trabalhos do Prof. Cesar Lattes neste período. Extraído do livro A. Marques, *Cesar Lattes 70 Anos: A Nova Física Brasileira*, Rio de Janeiro, CBPF, 1994, pp. 188-190, onde o leitor pode encontrar a relação completa dos trabalhos do Prof. Lattes.

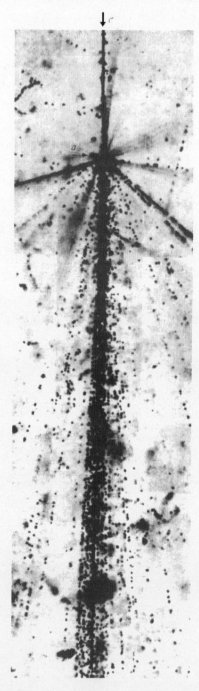

Figura 15.23: Produção múltipla de mésons. Extraída de C.F. Powell, P.H. Fowler and D.H. Perkins, *op. cit.*, p. 629.

4. C.M.G. Lattes, M. Schönberg and W. Schutzer, "Classical theory of charged point-particles with dipole moments", *Anais da Academia Brasileira de Ciências,* 1947; 19, pp. 193-245.

5. C.M.G. Lattes, P.H. Fowler and P. Cuer, "A study of the nuclear transmutations of light elements by the photographic method", *Proc. Phys. Soc.,* 1947; 59: pp. 883-900.

6. C.M.G. Lattes, P.H. Fowler and P. Cuer, "Range-energy relation for protons and α-particles in the new Ilford "nuclear research" emulsions", *Nature,* 1947; 159: pp. 301-302.

7. C.M.G. Lattes, G.P.S. Occhialini and C.F. Powell, "A determination of the ratio of the masses of π- and μ-mesons by the method of grain-counting", *Proc. Royal Phys. Soc.,* 1948; 61: pp. 173-183.

8. E. Gardner and C.M.G. Lattes, "Production of mesons by the 184-inch Berkeley cyclotron", *Science,* 1948; 107, pp. 270-271.

9. J. Burferin, E. Gardner and C.M.G. Lattes, "Positive mesons produced by the 184-inch Berkeley cyclotron", *Phys. Rev.,* 1949; 75: pp. 382-387.

10. H.L. Anderson and C.M.G. Lattes, "Search for the electronic decay of positive pions", *Nuovo Cim.,* 1957; 6: pp. 1356-1381.

11. C.M.G. Lattes and P.S. Freier, "Angular correlation between pions and muons measured in nuclear emulsions", *Proc. of the Padova Conf.,* 1957; 5: p. 17.

12. P.H. Fowler, P.S. Freier, C.M.G. Lattes, E.P. Ney and S.J.St. Lorant, "Angular correlation in the $\pi-\mu-e$ decay of cosmic ray mesons", *Nuovo Cim.,* 1957; 6: pp. 63-68.

13. P.H. Fowler, P.S. Freier, C.M.G. Lattes, E.P. Ney and D.H. Perkins, "A cosmic ray jet in the 1015 eV energy range", reported at *Varenna Int. Conf. on Cosmic Radiation,* June 1957.

16

Formación del Laboratorio de Física Cósmica de Chacaltaya

C. Aguirre Bastos[1]

Prólogo

Países pequeños como Bolivia tienen enormes limitaciones para generar nuevo conocimiento en la frontera de la ciencia. Las causas son muchas y existe una serie de diagnósticos que las identifican. La falta de recursos de infraestructura, humanos y financieros, el aislamiento de sus comunidades científicas son apenas algunas de las más evidentes. El atraso científico es, a su vez, motivo de atraso tecnológico y por tanto económico y social. En general, en estos países, el ambiente bajo el cual se intenta desarrollar la ciencia es extremamente adverso.

A pesar de las limitaciones, existen instituciones que, superando las condiciones del entorno han logrado realizar contribuciones de trascendencia al avance del conocimiento científico mundial. Uno de estos casos es el del Laboratorio de Física Cósmica de Chacaltaya de la Universidad Mayor de San Andrés de La Paz, cuya historia es el propósito de este relato.

---

[1] *N.E.*: Extraído de "Medio Siglo de Ciencia en Bolivia", Fundación Universitaria Simon I. Patiño, La Paz, Mayo de 1996. Trata-se de relato eloquente e muito completo sobre a história do Laboratório de Física Cósmica de Chacaltaya, que nossas limitações de espaço não permitem reproduzir na íntegra. Carlos Aguirre B. é Presidente da Academia Nacional de Ciências da Bolívia.

He considerado importante escribir un ensayo sobre el Laboratorio por varios motivos. El primero, "celebrar" medio siglo de aportes a la ciencia por parte de sus investigadores, la mayoría aún activos en cargos de diferente naturaleza. También recolectar la memoria personal y colectiva para dejar una historia documentada de esta institución que, sin duda ha sido una de las más importantes que ha tenido Bolivia, América Latina y el mundo durante años.

Al mismo tiempo, quise reconocer a quienes hicieron posible este dramático capítulo de la ciencia. Particular mención corresponde a los trabajadores "silenciosos", aquellos cuya dedicación hizo posible montar la infraestructura del Laboratorio. Son ellos los participantes claves de la "aventura" de Chacaltaya. Hubo muchos, acá apenas menciono a los que trabajaron durante el tiempo de mi permanencia en la institución: los hermanos José y Eusebio Monasterios, Gregorio Huayba, Raimundo Condori, Silverio Monasterios, Roberto Huanto, Mariano Huanto, Celestino Quisbert, Guillermo Mamani, Juan Bernal, Juan Calle, Samuel Beltrán, Elvira de Diaz, Angel Pabón y Andrés Villafán.

Junto a ellos, corresponde mencionar el aporte del personal administrativo-técnico principal: Jorge Saavedra, Antonio Carrasco, Jorge Bustos, Mario Ballivián, Teresa de Torrelio, Elsa Maceda, Gloria Pinel, Yolanda Escalante, Marilin Centellas de Trepp, Dardo Veramendi, Carlos Ormachea, Ivan Geier, Raúl Bascopé, Manuel de la Torre, Macedonio Trujillo, Lino Maldonado y Rodolfo Zalles. Es particularmente grato recordar a Ricardo Escobar Vallejo.

Este ensayo, que cubre un período de cincuenta años, puede contener algunas imprecisiones, fruto de la ausencia de alguna documentación o de la recolección "subjetiva" de hechos que se sucedieron en el Laboratorio durante mis 34 años de actividad junto a él. En todo caso, cualquier error es, naturalmente, de mi responsabilidad.

Aunque hubiese sido mi deseo consultar a muchos más actores de la "aventura" de Chacaltaya, razones de tiempo no lo permitieron. Quiero reconocer a quienes hicieron las contribuciones más importantes. El Profesor Ismael Escobar Vallejo escribió en marzo de 1993, una primera historia del Laboratorio, con motivo de recibir el Premio Wheatley de la American Physical Society por sus contribuciones al desarrollo de la ciencia boliviana. Posteriormente me escribió extensas cartas. El Ing. Ramón H. Schulczewski, rememoró en un corto ensayo, en 1987, sus vivencias y las de un grupo de jóvenes estudiantes que participaron activamente en los diferentes proyectos del Laboratorio.

Los Ings. Magín Zubieta, Ricardo Anda, José Antonio Zelaya y Ramón H. Schulczewski, hicieron valiosos comentarios y me facilitaron algunas fotografías. El Dr. Edison Shibuya de la Universidad de Campinas recolectó dos importantes trabajos producidos en Brasil: el de Alfredo Marques sobre la historia del descubrimiento del mesón pi y el de Yoichi Fujimoto sobre los orígenes de la Colaboración Brasil-Japón. El próprio Dr. Marques me envió su documento historia sobre el Centro Brasileiro de Pesquisas Físicas y varias fotografías.

Agradezco el aliento y la colaboración recibida durante la preparación de este ensayo a mis antiguos discípulos, el Lic. Alfonso Velarde Ch., ex-Director del Instituto de Investigaciones Físicas y el Dr. Miguel Peñafield, Decano de la Facultad de Ciencias Básicas de la Universidad Mayor de San Andrés. Agradezco, asimismo, la colaboración del distinguido periodista y amigo Armando Mariaca, por su paciente revisión del texto y valiosos comentarios.

Los Rayos Cósmicos

Como muchos fenómenos, la presencia de los rayos cósmicos fue detectada como consecuencia de otros estudios.[2] En sus investigaciones sobre la electricidad atmosférica Elster[3] y Geitel[4] notaron la existencia de una fuente no conocida de iones en el aire. Independientemente, Wilson en sus trabajos con una cámara de ionización sospechó de la presencia de un agente que penetraría grandes profundidades. El especuló que la fuente de esta radiación podría tener un origen extraterrestre.[5]

En el año 1911, el físico austríaco Victor Hess instaló, en un globo aerostático, tres electroscopios utilizados entonces para medir la radiación emitida por sustancias radiactivas. Las observaciones de Hess en este histórico vuelo fueron resumidas así: "Los resultados de mis investigaciones se explican por el supuesto de que una radiación de gran poder de penetración se introduce en nuestra atmósfera desde arriba".[6] Trabajos independientes de Kolhoster llegaron a esta misma conclusión. Estos descubrimientos dieron origen a una nueva disciplina de la física: la radi-

---

[2] Para una introducción a la Física Cósmica ver p.e.: G.R. Mejía y C. Aguirre "La Radiación Cósmica", Monografía No. 9, Serie de Física, Programa Regional de Desarrollo Científico y Tecnológico de la Organización de los Estados Americanos, Washington D.C. 1973. Para una descripción más avanzada ver S. Hayakawa "Cosmic Ray Physics", Monographs and Texts in Physics and Astronomy Vol. XXII, Wiley-Interscience, 1969. También T.K. Geisser "Cosmic Rays and Particle Physics", Cambridge University Press, 1990. Artículos interesantes sobre la historia de los rayos cósmicos son: Y. Sekido and H. Elliot, eds., "Early History of Cosmic Ray Studies" (Dordrecht, 1984); Q. Xu and L.M. Brown, "The Early History of Cosmic Ray Research", *American J. Phys.* **55** (1987) pp. 23-33.

[3] Elster J., *Phys. Zeits.* **2**, 560 (1900).

[4] Geitel H., *Phys. Zeits.* **2**, 116 (1900).

[5] Wilson C.T.R., *Proc. Camb. Phil. Soc.* **11**, 52 (1900) y *Proc. Roy Soc.(London)* **A68**, 151; **A69**, 277 (1901).

[6] Hess V.F., *Phys. Zeits.* **12**, 998 (1911) y *Phys. Zeits.* **13**, 1084 (1912). Muchos años después, en 1936, Hess recibió el Premio Nobel de Física "por el descubrimiento de la radiación cósmica". Este Premio lo compartió con el físico estadounidense Carl David Anderson "por su descubrimiento del positrón".

ación cósmica.

Ziegler[7] hace un interesante relato de la tecnología de los globos y el instrumental utilizado en ellos y su aporte al descubrimiento de los rayos cósmicos. En él menciona, con detalle, los experimentos de Wilson, Hess, Millikan, Wulf, Göckel y otros, incluyendo las disputas de Hess y Kolhoster con Millikan, originadas en los resultados de las mediciones de la intensidad de la radiación y principalmente en la prioridad del descubrimiento, que fuera resuelta por los vuelos realizados recién en 1932 por el francés Piccard,[8] que demostró lo correcto de las mediciones de Hess en los años 10. A través de sus trabajos, Millikan, fue el que popularizó la denominación de rayos cósmicos que sustituyó el término empleado hasta entonces en la literatura, "Höhenstrahlung" (radiación de altura).

Durante la década posterior a la Primera Guerra Mundial, la investigación sobre la radiación se caracterizó por dos acontecimientos importantes. Entre 1922 y 1927, Millikan descubrió la existencia de una radiación penetrante, en estudios realizados bajo agua en lagos de montaña, cuya profundidad correspondía a un espesor mayor del que normalmente impedía la penetración de rayos gama de la mayor energía conocidos entonces. Entre 1927-29, Clay demostró que la intensidad de la radiación cósmica disminuía al acercarse al ecuador y explicó esta variación como un efecto del campo magnético terrestre sobre la radiación, concluyendo que la componente primaria de ésta contenía, por lo menos en parte, partículas con carga eléctrica.[9]

Durante este período se desarrollaron nuevas técnicas de observación y detección de la radiación cósmica. En 1927 Skobelzyn adaptó una cámara de niebla con un campo magnético asociado al estudio de estas partículas. La invención y uso de outro instrumento – el llamado tubo Geiger-Müller – permitió detectar el paso de partículas cargadas (Bothe y Kolhoster) y el empleo de varios de estos tubos dispuestos geométricamente de manera apropiada, vino a constituir lo que se denominó un telescopio de rayos cósmicos.

Con el desarrollo de estas y otras técnicas, y a través de la observación de rayos cósmicos, Anderson pudo establecer la existencia del electrón positivo (positrón), postulado por la teoría cuántica de Dirac. En 1937, Nedermeyer y Anderson[10] mostraron que parte de la componente penetrante de la radiación cósmica era constituida por partículas con masas intermedias entre las del electrón y del protón siendo desde entonces conocidas como mesones. Ese mismo año Street y

---

[7] Ziegler C. A., "Technology and the Process of Scientific Discovery: The Case of Cosmic Rays", *Technology and Culture* Vol. **34**, No. 4 October 1989, pp. 939-963.

[8] Piccard A., "Kosmische Strahlung", en Prof. Piccard's *Forshungsflug in die Stratosphäre*, Naumann y Hohenester eds. Ausburg 1931.

[9] Clay J., *Proc. Amsterdam* **30**, 1115 (1927).

[10] Nedermeyer S.H. and Anderson C.D., *Phys. Rev.* **51**, 884 (1937).

Stevenson[11] fueron capaces de determinar la masa del mesón igual a 175 veces la masa del electrón (m_e), valor que luego de posteriores análisis pasó a 200 m_e.

Todos estos resultados experimentales tuvieron un impacto importante sobre el desarrollo de las teorías de la electrodinámica cuántica que eran investigadas teóricamente por físicos como Bohr, Compton, Klein, Nishina, Dirac, Bethe y otros.

En este período, dos nuevas técnicas – el método de coincidencias (Rossi) y la expansión controlada de la cámara de niebla (Blackett y Occhialini) – condujeron a un avance importante en el conocimiento de la radiación cósmica secundaria; es decir, una cascata de partículas, que se produce en la atmósfera de la tierra como consecuencia del choque de una partícula extra terrestre (el rayo cósmico primario) con un nucleo atmosférico[12] (e.g. un "chubasco atmosférico extenso"). Por medio de un detector de chubascos Rossi observó su frecuencia en función del espesor de un material productor. La curva frecuencia-espesor permitió concluir que un chubasco era un fenómeno que involucraba electrones y posiblemente fotones. En 1936 Pfötzer[13] observó variaciones de intensidad de la radiación con la altura y estableció que el máximo de intensidad ocurre a los 15 km sobre el nivel del mar. El descubrimiento mismo de los chubascos, sin embargo, corresponde al físico francés Pierre Auger.

En 1941 Rasetti[14] detectó una fracción inestable en la componente penetrante, con lo que mostró la presencia de una actividad bajo la forma de emisión de partículas ionizantes, atrasada con relación al paso del grupo penetrante y decayendo con una vida media de 2 microsegundos. Estas observaciones fueron seguidas por excepcionales experimentos ejecutados por Conversi, Pancini y Piccioni, mostrando que partículas en la componente penetrante, con cargas eléctricas opuestas, se comportaban de manera diferente.

Actualmente, se conoce que los "chubascos atmosféricos extensos" están compuestos por una componente "suave" (electrones-sus antipartículas-fotones), otra "dura" (mesones mu – positivos y negativos), una nuclearmente activa (hadrónica: protones, neutrones, piones, kaones, hiperones, etc. y sus antipartículas correspondientes) y una neutrínica (neutrinos).

El estudio de los chubascos ha sido tradicionalmente dividido en dos partes: el desarrollo longitudinal a lo largo de la atmósfera y la estructura lateral a diferentes puntos fijos de la atmósfera. Los diferentes problemas derivados del complejo fenómeno de cascadas en la atmósfera fueron tratados por muchos autores. La evidencia experimental que se acumuló en los años 1920, dió lugar a la teoría de cascadas desarrollada por Bhabha y Heitler y por Carlson y Oppenheimer.

---

[11] Street J.C. and Stevenson E.C., *Phys Rev.* **52**, 1003 (1937).

[12] Rossi B., *Z. Physik* **82**, 151 (1933).

[13] Pfötzer G., *Z. Physik* **102**, 23 (1936).

[14] Rasetti F., *Phys. Rev.* **59**, 706 (1941).

Los primeros cálculos sobre las cascadas fueron hechos por Euler y Wergeland[15]. La llamada función de distribución lateral fue obtenida por Molière[16] y es aún utilizada para efectos de comparación con los resultados experimentales, siendo aplicable solamente a las cascadas cerca de su máximo desarrollo. Frente a esta limitante, Nishimura y Kamata[17] extendieron su análisis a todas las etapas de desarrollo de la cascada en la atmósfera. Greisen, a su vez, introdujo algunas modificaciones.[18]

Uno de los resultados importantes del análisis es el establecimiento de una ecuación que permite establecer la energía del rayo cósmico primario en función del número máximo de partículas (N_{max}) detectadas en un cierto nivel de observación. Tal expresión está dada por:[19]

$$E = 1.9 \times 10^9 N_{max} \text{ eV}$$

Otros parámetros que pueden ser establecidos por dispositivos de detección de chubascos son la posición del eje (alrededor del cual se concentra un número importante de las partículas), la curvatura del frente, la edad (una medida del grado de desarrollo del chubasco en la atmósfera), la dirección de arribo, etc.

Durante muchos años y a partir de la década de los 30, se estudiaron con profundidad los efectos de modulación de los campos magnéticos terrestre, solar, galáctico o extragaláctico sobre los rayos cósmicos. El estudio de los aspectos geofísicos de la radiación cósmica se refiere al comportamiento de las partículas dentro de los campos magnéticos terrestre y solar. Los estudios pioneros de Lemaître y Vallarta mostraron que el cielo puede dividirse en cuatro partes, aquella en que los rayos cósmicos de cierta energía no son permitidos a entrar en la atmósfera, otra donde lo son, y las otras dos partes de penumbra y de exclusión por el horizonte de observación.

Se puede definir la resistencia de la partícula a los efectos de desviación del campo por la relación:

---

[15] Euler H., Wergeland H., *Astrophys. Norveg* **3**, 1940 p. 165 y *Naturwissenschaften* **28**, 1940 p. 41

[16] Molière G., *Cosmic Radiation* (W. Heisenberg ed.), New York, Dover, 1946.

[17] Nishimura J. and Kamata K., *Prog. Theo. Phys., Osaka* **5** (1950) p. 899; **6** (1951) p. 628; **7** (1952) p. 185.

[18] Greisen K., *Progress Cosmic Ray Physics*, vol. III (J.Wilson ed.), North-Holland, Amsterdam, 1956.

[19] A lo largo de este ensayo se utiliza el término electrón voltio (eV) como medida de energía, siguiendo la práctica en la física cósmica. En unidades térmicas 1 eV es equivalente a 3.829×10^{-20} calorías. Se debe recordar que una cantidad de energía diaria apropiada para el desarrollo del cuerpo humano es del orden de una 3 a 4 000 calorías, aproximadamente 10^{22} eV, que es algo más de la energía mas alta de una partícula de la radiación cósmica primaria observada hasta ahora. En términos eléctricos 1 eV es igual a 4.451×10^{-26} Kw-hr o 1.602×10^{-19} joules. Una partícula cósmica primaria de muy alta energía se acerca así a 1 Kw-hr.

$$H r = p/Z e,$$

donde H es el campo magnético, r es el radio de curvatura de la partícula y p/Ze la resistencia. El parámetro p es una función de la posición geomagnética y explica la variación de intensidad de los rayos cósmicos con la latitud geográfica, la que aumenta desde el ecuador hacia los polos.

Los estudios sobre la variación de intensidad de los rayos cósmicos mostraron la existencia de cuatro categorías, dependientes de las variaciones en el tiempo del campo magnético terrestre: (1) de 11 años, correspondiente al ciclo solar; (2) de 27 días, correspondiente a la rotación del Sol en torno a su eje; (3) diurnas y semidiurnas, que corresponden a la rotación de la Tierra en torno del Sol y (4) hechos de corta duración originados por la actividad solar (por ejemplo una tormenta eléctrica).[20] Las variaciones temporales de los rayos cósmicos están sujetas al llamado "viento solar", un plasma de partículas ionizadas que adquiere velocidades de cientos de km/h.

En 1947 se descubrió en Chacaltaya, Bolivia, el mesón pi (una nueva partícula) y el decaimiento de éste en méson mu (la partícula de Anderson, Nedermeyer, Street y Stevenson), hecho de enorme trascendencia en la física nuclear. Importantes resultados fueron también obtenidos en este Laboratorio, en la década de los años 1960, con las observaciones realizadas por la colaboración Brasil-Japón; entre ellos se estableció la existencia de interacciones nucleares exóticas, que hoy son objeto de estudio tanto en nuevos experimentos con rayos cósmicos como con aceleradores artificiales de partículas.

Un parámetro importante en el estudio de los rayos cósmicos es la intensidad, expresada por el número de partículas que, por unidad de tiempo, arriban a una superfície dada, bajo condiciones geométricas específicas. La distribución de intensidad es representada por un espectro. Los estudios desarrollados por diversos dispositivos de observación han permitido establecer por lo menos tres grandes porciones en el espectro de rayos cósmicos primarios. La primera a energías entre 10^{13} eV y 10^{15} eV, la segunda entre 10^{15} eV y 10^{17} eV y la tercera a energías superiores a 10^{17} eV.[21] En el rango intermedio, la forma del espectro fue establecida por mediciones hechas en Chacaltaya.

En 1948 se encontró que la radiación cósmica primaria contiene núcleos pesados además de protones (Freier y colaboradores). Este descubrimiento tuvo importantes implicaciones en la astrofísica, puesto que mostró que no existen procesos violentos en el universo que podrían destruir nucleos en el curso de

---

[20] Designados como fenómenos Forbush en honor a su descubridor.

[21] Algunas observaciones más recientes permiten cuestionar esta visión clásica del espectro de rayos cósmicos.

su aceleración. En general, en este período, se pudieron establecer importantes resultados sobre la composición de la radiación. A bajas energías ésta fue determinada directamente volando detectores en globos. A medida que la energía aumenta, la composición es determinada por métodos indirectos.

El cuadro 1 muestra la abundancia relativa de la componente nuclear en la región de energía de 0.5 GeV a 10^3 GeV por nucleo.[22]

Nucleo	Abundancia en la R.C.	Abundancia Universal
H	1.5×10^3	0.7×10^3
He	100	100
Li,Be,B	1.6	10^{-5}
C,N,O,F	6.7	1
Z>10	25	0.1 a 0.4
Z>20	0.7	0.005 a 0.02

Además de nucleos existen isótopos, lo que sugiere que la fragmentación de nucleos pesados en el espacio juega un papel importante en la composición observada. La radiación contiene, al mismo tiempo, electrones de baja energía, producidos por colisiones entre protones en el espacio. El espectro de energía de tales partículas está bien definido entre 10^6 y 10^{11-12} eV. Existen también evidencias de electrones galácticos producidos y acelerados en las fuentes. Algunas fuentes puntuales han sido detectadas, tal el caso de VELA X. Finalmente, existen en la radiación primaria rayos gama de baja energía. A energías mayores, en particular por encima de 10^{17} eV, la evidencia experimental respecto a la existencia de rayos gama primarios, no es suficiente para dar una respuesta satisfactoria.

En los años 1950 se desarrollaron nuevas teorías sobre el origen de los rayos cósmicos. En particular la existencia de campos magnéticos celestes fue enfatizada por Alfven.[23] Anteriormente, en 1933, Swann había propuesto la aceleración de rayos cósmicos por campos magnéticos variables. Una propuesta de origen solar fue adelantada por Richtmeyer y Teller y el origen galáctico por Fermi. En los años

---

[22] 1 GeV = 10^9 eV.

[23] Alfvén H., *Cosmical Electrodynamics*, (1950).

1960 Ginzburg y Hayakawa hicieron importantes contribuciones al discutir los conceptos de origen extragaláctico junto a los orígenes galáctico y solar.

Más recientemente, a energías muy altas, entre 2×10^{18} eV y 10^{19} eV, se ha podido construir una visión que señala una anisotropía galáctica concentrada. En términos de un modelo simple, es posible que más del 50% de los rayos cósmicos provengan de fuentes asociadas con el plano galáctico a energías del orden de 10^{19} eV. A energías mayores existe evidencia de un origen extragaláctico con sugerencia de que las fuentes de creación de estas partículas se encuentran en el Superconglomerado de Virgo.[24] La verificación de estas hipótesis sólo será posible con dispositivos más perfeccionados. Actualmente existen dos propuestas para el final de los años 90. La primera, hecha por un grupo liderado por el Premio Nobel de Física de 1980 James Cronin, de construir un dispositivo de chubascos atmosféricos extensos que cubra 5 000 km^2, y la segunda, hecha por el grupo de Chacaltaya,[25] de construir un dispositivo que combine emulsiones nucleares con equipos electrónicos usados para la detección de chubascos. Estas son dos propuestas complementarias en la práctica.

En 1957, com la ayuda de satélites artificiales, Van Allen y Vernov descubrieron, independientemente, la existencia de anillos o bandas de partículas atrapadas en el campo magnético terrestre. En 1961, se detectaron electrones primarios y, en 1962, rayos X y sus fuentes de origen (Giacconi y colaboradores). De importancia para los estudios de los rayos cósmicos fue el descubrimiento de la radiación de fondo de 2.7 ^{o}K, interpretada como remanente del desacoplamiento de la materia y la energía en las tempranas etapas del universo (durante el "big-bang").

En 1965 se observó por primera vez la interacción de un neutrino (la partícula más "pequeña" existente en el universo) con otras partículas (Reines y Cowan). Posteriormente,en las décadas de los años 60 y 70, grandes dispositivos experimentales[26] para observación de chubascos atmosféricos extensos, permitieron determinar el espectro de energía de rayos cósmicos muy energéticos, facilitaron información acerca de la composición química y permitieron avanzar algunas hipótesis sobre su origen y sobre las interacciones nucleares de alta energía, en particular, sobre la existencia de nuevas partículas componentes de la materia.

Los años 1980 vieron la integración de partes de la física cósmica con la astrofísica, la geofísica, la física del espacio y otras disciplinas. La existencia de aceleradores artificiales de partículas que pueden detectar un número mucho

---

[24] Watson A. A., "The Highest Energy Cosmic Rays", *Nuclear Phys. (Proc. Suppl.)* **22B** (1991) 116-137, North-Holland.

[25] Aguirre C. *et al*, "Experiment OMEGA: Observation of Multiple Particle Production, Exotic Interactions, Gamma Ray Air Showers and Heavy Primaries", Cosmic Ray Research Laboratory University of Tokyo 1989.

[26] Volcano Ranch (1959-1963); Chacaltaya (1962 al presente); Haverah Park (1968-1987); Sydney (1968-1979); Yakutsk (1974 al presente); Akeno (1979 al presente); Utah (1981 al presente).

mayor de eventos que aquellos con los rayos cósmicos, obligaron a la física cósmica a subir el límite de energía para sus observaciones. En este tránsito, la física cósmica está dedicando hoy un esfuerzo especial al estudio de fenómenos de las más altas energías existentes en el universo.

Es precisamente en los años posteriores a 1950, que el Laboratorio de Física Cósmica de Chacaltaya, empieza a trabajar y producir importantes resultados, que han contribuido al avance de esta ciencia. Antes de Chacaltaya, los científicos cósmicos tenían a su disposición facilidades instaladas en Pic du Midi en los Pirineos, Jungfraujoch en los Alpes o Aiguille du Midi, en Chamonix, todos a alturas entre 3 000 y 3 500 metros sobre el nivel del mar.

Chacaltaya disfruta de una situación excepcional por diversos motivos: la principal, su altura entre 5 600 m (el pico más alto) y 5 200 m (donde se encuentran las instalaciones principales del Laboratorio), en la cual se produce el máximo desarrollo de los chubascos atmosféricos extensos producidos por primarios del orden de los 10^{14-17} eV. Otra ventaja es su posición justo al "frente" del centro de la galaxia. Al mismo tiempo, Chacaltaya está situada sobre el ecuador geomagnético y, por tanto, en el lugar ideal para estudiar las variaciones temporales de la radiación cósmica y observar fuentes puntuales de rayos cósmicos en ambos hemisferios. La altura, al mismo tiempo, permite la reducción de los efectos secundarios, dada la pequeña espesura de la capa atmosférica (aproximadamente mitad de la presión del nivel del mar). Finalmente, su fácil acceso es también ventaja importante.

Hoy se vislumbran nuevas posibilidades de investigación, referidas en particular a los aspectos astrofísicos y de física nuclear a energías muy altas. En este contexto, Chacaltaya continua teniendo ventajas comparativas sobresalientes sobre otros lugares de observación e instalaciones existentes.

La creación del Servicio Meteorológico de Bolivia

La historia de la física cósmica en Bolivia data de 1927, cuando una expedición del físico norteamericano Millikan, visitó América del Sur realizando observaciones sobre la variación de intensidad de la radiación cósmica con la latitud. Esta misión no involucró a científicos bolivianos[27] y no creó interés en continuar con las investigaciones en esta naciente disciplina.

La historia del Laboratorio se remonta a finales de la Guerra Civil española, durante la cual el Dr. Ismael Escobar Vallejo trabajaba para el servicio de meteo-

---

[27] Entre otros porque no existía en el país ninguna escuela de física en las universidades o fuera de ellas.

rología de la Fuerza Aérea de España. A inicios de 1939, Escobar llegó a Francia como refugiado político y a finales de ese año – muy poco antes de la caída de París – y en uno de los últimos trenes que salieron de esa ciudad, consiguió tomar un barco francés que se dirigía a la isla de Martinica y que lo dejó en Santo Domingo. Después de un año en la República Dominicana, pasó a Bolivia en julio de 1941.

Escobar empezó trabajando como meteorólogo en enero de 1942. Su trabajo inicial consistió en montar y operar una red de estaciones para apoyar los trabajos del Ministerio de Agricultura.[28] Fue Jefe de Sector en el Comité Fiscal de Fomento Agrícola y Regadío, dentro el cual preparó el borrador de decreto que posteriormente crearía el Servicio Meteorológico Boliviano. Este Comité dependía del Ministerio de Economía y lo dirigía el ingeniero mexicano Eligio Esquivel, que años después sería Gobernador del Estado de Baja California. En este período el Presidente de la República era el General Enrique Peñaranda y el Ministro de Economía don Alberto Crespo Gutierrez. En el Comité, Escobar fue colega de Hugo Mansilla Romero, quien tendría posteriormente tareas de responsabilidad en la conducción de la Universidad por muchos años.

En el primer informe de Escobar al Director General de Agricultura, correspondiente a los meses de febrero a marzo de 1942, después de la carta de presentación, se encuentra el ante proyecto de decreto supremo creando el Servicio.[29] En fecha 21 de diciembre de 1942 recibió el memorandum 159/334/42 firmado por el Ing. Eduardo Palomo S., contratándolo como Jefe del Servicio Meteorológico de Bolivia, dependiente de la Dirección General de Agricultura. Posteriormente este cargo se elevó a la categoría de Director General.[30]

La primera Estación Central de la red meteorológica se instaló en la Escuela Militar de Aviación en El Alto. Escobar fue desde marzo de 1942 y hasta el traslado de la Escuela a Santa Cruz, profesor de meteorología. Ello le permitió hacer sondeos de la atmósfera superior con un bimotor Junker, hasta cerca de los 7 000 metros.

---

[28] Parte de los primeros subtítulos de este ensayo han sido tomados de Ismael Escobar ante la American Physical Society, en marzo de 1993. Otras notas han surgido del exámen de los extensos archivos del Laboratorio y notas complementarias recolectadas en conversaciones sostenidas con Escobar y Cesar Lattes. Varias notas fueron complementadas a partir de las cartas del primero, dirigidas al autor durante 1994 y 1995.

[29] El informe obra en poder de Ismael Escobar.

[30] No sin antes tener Escobar una gran "trifulca" con el entonces ministro de Agricultura, General Noriega, quién designó a un capitán como asistente, encargado de controlar la firma de un libro de entradas y salidas. Escobar intentó convencer al ministro, inclusive con amenazas de renuncia, que las horas de trabajo en el Servicio debían corresponder a las intenciones de recepción radial, no siempre coincidentes con las oficiales. Se resolvió el problema, firmando el capitán por Escobar las entradas y salidas formales y trabajando este último entre diez y doce horas, muchas fuera del horario oficial.

En septiembre de 1942, se instaló en el Monte Chacaltaya, a 5 200 metros, una de las estaciones de la red meteorológica. Esta estación incluyó equipos para medir el tiempo e intensidad de la radiación solar. La instalación de la estación se hizo factible por la existencia de un camino y de una casa construida dos años antes por el Club Andino Boliviano. Participaron en esta tarea Raul Posnanski y Alfredo Hendel, quien estuvo vinculado al Laboratorio desde sus inicios y es actualmente Profesor Emérito de la Universidad de Michigan. Escobar llegó a ser Secretario del Club y en 1943, con el Ministro de Educación Rodas Eguino, se dio el primer curso de esquí, a estudiantes y aficionados de La Paz.

El 4 de julio de 1942 el gobierno de Enrique Peñaranda dictó un decreto supremo que declaró Parque Nacional al grupo andino comprendido entre los pasos Condoriri y Yungas: Huayna Potosí, Chacaltaya y Laguna Milluni.

Lo que se concibió en 1942 como una red de estaciones para "monitorear" las condiciones del tiempo, se convirtió, gracias al impulso de Ismael Escobar en un eficiente Servicio Meteorológico de Bolivia. En 1943 el Presidente de la República, Coronel Gualberto Villaroel, inauguró en la cima de Chacaltaya (5 600 m.s.n.m.) una nueva caseta meteorológica. A partir de este año se inician observaciones con importantes aportes al conocimiento de la climatología local. Los informes meteorológicos mejoraron el análisis de la dinámica de las masas de aire y, consecuentemente, la precisión de las medidas y previsiones de tiempo para el aeropuerto y la ciudad de La Paz. Se inició también la publicación de la importante revista científica "Nimbus".[31]

El 19 de junio de 1943, mediante Oficio 54/953, la Universidad Mayor de San Andrés designó a Escobar Profesor Interino de la Escuela de Ayudantes Técnicos en la materia de Cosmografía y Meteorología. Más tarde, en septiembre de 1947, mediante memorandum 193/947, fue designado Catedrático de Complementos de Física en el primer curso y de Cosmografía y Meteorología en el segundo, del Instituto de Ciencias Exactas. Esta designación fue el resultado de un examen de oposición. Como señala el propio Escobar, "la UMSA, a través de uno de sus profesores, estuvo integrada a los trabajos de Física y Termodinámica de la atmósfera, incluyendo radiaciones solares y cósmicas, mucho antes de oficializarse la creación de Chacaltaya, como Laboratorio..."

Punto de Partida de la Gran Aventura de Chacaltaya

Al finalizar la Segunda Guerra Mundial, el físico brasileño Cesar Lattes se encontraba trabajando en la Universidad de São Paulo con una cámara de

---

[31] La Biblioteca del Instituto de Investigaciones Físicas de la Universidad Mayor de San Andrés preserva la colección completa de esta revista.

niebla construida en colaboración con Ugo Camerini y Andrea Wataghin. Varias fotografías obtenidas con este equipo fueron enviadas a Giuseppe Occhialini que se había integrado al grupo de rayos cósmicos dirigido por Cecil Frank Powell en Bristol, Gran Bretaña. A manera de respuesta, Lattes recibió fotomicrografías de trazos de protones y partículas alfa obtenidas en un nuevo tipo de emulsión producida experimentalmente por la empresa Ilford. Estas motivaron al científico brasileño a solicitar incorporarse al grupo de Bristol, pues consideró que se había abierto un potencial inmenso en el estudio de los rayos cósmicos por métodos fotográficos. Fue así que llegó a Gran Bretaña en el invierno de 1945/46.[32]

Una vez en Bristol Occhialini y Lattes fueron de la idea de llevar las nuevas emulsiones nucleares a sitios elevados. Para identificarlos, Lattes visitó el Departamento de Geografía de la Universidad de Bristol y "descubrió" la estación meteorológica de Chacaltaya. Es también probable que, por sugerencia de los meteorólogos, Dr. Mariano Doporto Marchori, Director del Servicio Meteorológico de Irlanda y del Dr. Duperier, colaborador del laureado Nobel, Prof. Blackett, en Londres, quienes conocían Chacaltaya y a Ismael Escobar, que el grupo de Bristol se propone utilizar esta estación. Años más tarde, el propio Blackett sería invitado a Bolivia, para participar en los proyectos del laboratorio.

En 1946 Lattes llega a Bolivia, haciendo uso de sus vacaciones y prácticamente con sus propios recursos;[33] trae consigo las emulsiones nucleares que expone en Chacaltaya. Estas placas, al revelarse, muestran los trazos de 30 mesones dobles. Se decide entonces medir la relación de masas del primero y segundo mesón haciendo un repetido contaje de granos sobre los trazos. El resultado muestra de manera convincente que se trata de un nuevo proceso. Se identifica al mesón más pesado con la partícula prevista teóricamente por Yukawa en los años 1930 (partícula pi) y la secundaria con el mesotrón previsto por Anderson y otros (partícula mu).

El descubrimiento del mesón π y su decaimiento en mesón π, partículas constituyentes de la materia y responsables de las fuerzas nucleares, revela pronto la importancia de la técnica utilizada en el estudio de los fenómenos cósmicos y el lugar elegido. El descubrimiento hace al Profesor Hideki Yukawa del Japón acreedor al Premio Nobel de Física en 1949 y al Profesor Cecil F. Powell acreedor al mismo Premio en 1950.[34] Los trabajos científicos publicados en la prestigiosa revista *Nature* bajo la autoría de Lattes, Occhialini y Powell hacen conocer

---

[32] C.M.G. Lattes, "My work in Meson Physics with Nuclear Emulsion". N.E.: também neste volume.

[33] Marques A., "25 Anos da Descoberta do Meson Pi", Centro Brasileiro de Pesquisas Físicas, *Ciência e Sociedade* No. 12, Rio de Janeiro 1973.

[34] Las citas de la Fundación Nobel rezan: Al Prof. Hideki Yukawa de la Universidad Imperial de Kyoto del Japón y de la Universidad de Columbia de los Estados Unidos "por su predicción de la existencia de mesones sobre la base del trabajo teórico sobre las fuerzas nucleares". Al Prof. Cecil Franck Powell de la Universidad de Bristol, Gran Bretaña, por su "desarrollo del método".

Chacaltaya a todo el mundo.[35]

Hacia finales de 1947, Lattes deja Bristol para dirigirse a Berkeley en California con la intención de detectar en el ciclotrón de esa universidad, piones producidos artificialmente. Poco tiempo después, Gardner y Lattes realizan los descubrimientos de los mismos.[36] Ello confirmó plenamente las expectativas que el físico E.O. Lawrence y sus colaboradores pusieran sobre la máquina que desarrollaron, justificando su empresa y abriendo una nueva era científica, la de los aceleradores de partículas.[37]

Como señala Occhialini, el descubrimiento en Chacaltaya es parte importante de la história de la ciencia[38] y Cesar Lattes fue ciertamente uno de su grandes actores. Alfredo Marques, en su análisis, señala "para nosotros en el Brasil y América del Sur, el descubrimiento tuvo dos extensiones importantes: la revitalización y potenciación de los propósitos del viejo observatorio, hoy Laboratorio de Física Cósmica en Chacaltaya, vinculado a la Universidad Mayor de San Andrés y la creación en Rio de Janeiro, del Centro Brasileiro de Pesquisas Físicas".

El descubrimiento tuvo otros impactos, como lo atestigua Taketani,[39] al señalar que 1947 fue un año de gran importancia para la física de partículas elementales, especialmente en Japón. Permitió confirmar la teoría de Yukawa, la teoría de dos mesones de Sakata y la teoría de Tomonaga (quien posteriormente, en 1965, ganaría un Premio Nobel por sus trabajos en electrodinámica cuántica). Mostró a los físicos japoneses, quienes no tenían actividad en áreas de la física experimental por la destrucción del Japón, luego de la Guerra, que era posible trabajar en esa disciplina de la ciencia. A partir de entonces, se organizó un grupo de jóvenes científicos dedicados a los rayos cósmicos, al mismo tiempo que se incentivó a empresas japonesas (Fuji Film y Konishiroku Film) a producir emulsiones

---

[35] C.M.G. Lattes, G.P.S. Occhialini and C.F. Powell (H.H. Wills Physical Laboratory of the University of Bristol) "Observations on the Tracks of Slow Mesons in Photographic Emulsion Plates", *Nature* **160**, 453 (Oct., 4) and 486 (Oct. 11), 1947.

[36] Gardner E. and Lattes C.M.G., *Science* **109**, 270 (1948).

[37] El físico norteamericano H.L. Anderson señaló, en el Coloquio Internacional sobre la Historia de la Física de Partículas (París, Julio 1982), "el sincrociclotrón emitió su primer haz justo antes de la medianoche del 1 de novembre de 1946. Aunque los piones habían sido abundantemente creados, las tentativas para encontrarlos fallaron por ausencia de técnicas de emulsión apropiadas. Ellos fueron encontrados casi inmediatamente después que Lattes arribara de Bristol con la técnica y las emulsiones Ilford apropiadas. Lattes era el joven brasileño que, trabajando con Occhialini y Powell en Bristol, fue el primero en encontrar piones en los rayos cósmicos. Ahora los encontró producidos artificialmente en una máquina" (cita traducida por el autor) de una cita hecha por J. Leite Lopes en su artículo "O CBPF e a Nova Física no Brasil" em *Cesar Lattes 70 Anos: A Nova Física Brasileira*, A. Marques (Editor), CBPF, Rio de Janeiro, Diciembre 1994.

[38] Occhialini G., "Cesar Lattes: the Bristol Years en *Topics in Cosmic Rays*, J. Bellandi Filho y A. Pemmaraju (eds.), Universidade de Campinas, Campinas, SP, 1994. N.E.: Também publicado neste volume.

[39] Taketani M., "Professor Lattes e a Física do Japão", en *Topics in Cosmic Rays, op. cit.*

fotográficas de gran sensibilidad.

Primeros Experimentos en Chacaltaya y la Creación del Laboratorio en 1951

Con los antecedentes señalados, se estudió la posibilidad de crear una institución permanente que sirviera de base a la comunidad científica internacional para sus investigaciones sobre la radiación cósmica.

Al comienzo de 1949, Ismael Escobar presentó al Rector de la Universidad una propuesta para construir en Chacaltaya un Laboratorio.[40] Puesto que éste se localizaría en un Parque Nacional, la Universidad inició un trámite para obtener un permiso especial, que fue concedido poco tiempo después.[41] El decreto supremo 02921 de fecha 9 de enero de 1952 promulgado por el Presidente, General Hugo Ballivián, declaró de "necesidad y utilidad pública la construcción de un Observatorio de Física Cósmica en Chacaltaya".

Ese mismo año, Maurice Shapiro del Naval Research Laboratory de Washington, Herman Yagoda del National Bureau of Standards, Marcel Schein de la Universidad de Chicago y Hervasio de Carvalho del Centro Brasileiro de Pesquisas Físicas visitaron La Paz para exponer emulsiones fotográficas en Chacaltaya y otros sitios. Con la colaboración del Club Andino Boliviano un grupo consiguió colocar algunas de ellas en el monte Sajama, las que nunca pudieron ser recuperadas;[42] de Carvalho colocó emulsiones a diferentes profundidades del lago Titicaca.

En septiembre de 1950 Escobar obtuvo una beca de la John Simon Guggenheim Memorial Foundation y junto al Profesor Bruno Rossi inició la colaboración del Laboratorio con el Massachussetts Institute of Technology, la que duraría cerca de 20 años. En 1951 visitó la Universidad de Colorado (Echo Lake Laboratory) donde junto a Frank Harris probó nuevos equipos para determinar la llamada anisotropía Este-Oeste de la intensidad de mesones π.

Más tarde, en 1951, el Rector de la Universidad Mayor de San Andrés y el Consejo Universitario aprobaron oficialmente la creación del Laboratorio de Física Cósmica de Chacaltaya, como un "*centro de investigaciones, enseñanza y observaciones meteorológicas*" a ser operado en parte por la Universidad Mayor de San Andrés y en parte por el Servicio Meteorológico de Bolivia. En la creación

---

[40] I. Escobar V. "Anteproyecto para el Observatorio de Física Cósmica Chacaltaya", *Nimbus*, La Paz (1949).

[41] Prácticamente por el propio Escobar puesto que trabajava en el Ministerio de Agricultura, institución encargada de otorgar tales permisos.

[42] I. Escobar V., "Una Expedición al Sajama", *Nimbus* **4** y *Revista Metereológica del Uruguay*, Montevideo (1950).

del Laboratorio se debe reconocer la visión de las autoridades de la Universidad, entre ellas su Rector Claudio Sanjinés y los Profesores Vicente Burgaleta (Decano de la Facultad de Ingeniería hasta su muerte en 1952) – Jorge Muñoz Reyes y Hugo Mansilla Romero (Decano de la Facultad de Ingeniería en sustitución de Burgaleta) y quienes serían a su vez rectores en años posteriores. Además de ellos, muchísimas autoridades y profesores de la Universidad contribuyeron a la creación y posterior crecimiento y consolidación del Laboratorio.[43]

La Década de los Años 50

Las actividades oficiales del Laboratorio se inician el 1º de enero de 1952. Ismael Escobar es designado su primer Director.[44] Son figuras importantes en este primer período, los Rectores Claudio Sanjinés (hasta agosto de 1952) y su sucesor Pedro Valdivia.

La Universidad Mayor de San Andrés autorizó, el 21 de enero, la construcción de una cabaña de madera en Chacaltaya[45] y contribuyó con 5 510 pies de este material a este propósito.[46] Por outro lado, Escobar obtuvo gratuitamente una pequeña casa de aluminio, la que fue llevada a Chacaltaya por estudiantes, amigos y miembros del Club Andino el día 3 de marzo.[47] Esta sirvió como el primer albergue para los científicos. Al mismo tiempo, se acopló un motor Diesel para proveer la energía eléctrica necesaria a los equipos que ya habían sido probados en Echo Lake para el estudio de la asimetría de mesones positivos y negativos en el ecuador geomagnético. En agosto de este año se inauguró la primera construcción.[48]

Frank Harris y su familia llegaron a La Paz en abril de 1952, justo a tiempo para pasar varios días en el sótano de su casa, mientras fuera tenía lugar una de las más sangrientas revoluciones de la historia de Bolivia. Él se constituyó en el primer investigador visitante oficial del recientemente creado Laboratorio.[49] El experimento de Harris culminó con su tesis doctoral. Esta trató sobre el fenómeno de la intensidad de muones y fue dirigida por Rossi en el MIT, quien, de acuerdo

---

[43] Aunque hubieron también muchos detractores.

[44] Memorandum del Rector Claudio Sanjinés No. 12/18240/23316 de fecha 18 de enero de 1952.

[45] Oficio del Rector Sanjinés No 12-1842-23330

[46] El Ingeniero Hugo Zárate también Rector de la Universidad en años posteriores, realizó los primeros diseños de las edificaciones de Chacaltaya.

[47] Agradezco al Ing. David Tejada por este dato. El Ing. Tejada y el Dr. Alfredo Hendel fueron contratados por la Universidad a requerimiento de Ismael Escobar el 7 de diciembre de 1951. Fueron dos de los primeros funcionarios de la institución. En los primeros meses de 1952 se integran también al Laboratorio Rafael Vidaurre y Oscar Troncoso.

[48] Por este motivo, el mes de agosto ha sido tomado equivocadamente por varios funcionarios del Laboratorio, en sus discursos y trabajos, como la fecha de fundación del mismo.

[49] Designación del Rector de la Universidad mediante Oficio 12-18490-23663 de fecha 10 de abril de 1952 como "Ayudante Interino".

al propio autor, le dió más de un dolor de cabeza por su meticulosidad en su revisión.[50] Entre muchos otros comentarios interesantes que, en el transcurso de su preparación hacía, Harris indicaba "estoy seguro que no hay más del 10% de partículas positivas entre la radiación primaria". Los resultados finales de la tesis sobre las variaciones de la intensidad de muones, fueron resumidos en un importante e histórico artículo publicado en el Physical Review bajo la autoría de Escobar y Harris.

Entre 1951 y 1953 Escobar cumplió la doble función de Director del Servicio Meteorológico de Bolivia y Director del Laboratorio de Física Cósmica. El 19 de diciembre de 1952, presentó su renuncia al Servicio. El Ministro de Agricultura, Germán Vera Tapia, en su aceptación mediante nota D69/53 de fecha 2 de enero de 1953, resalta la importante contribución de Escobar en la creación y fortalecimiento de esta institución. A tiempo de dejarla, asumió las funciones de profesor en el Centro Brasileiro de Pesquisas Físicas.

Fue también en estos años que Juan Hersil, graduado en el MIT como ingeniero, trabajó con George Clark y Bruno Rossi preparando un experimento sobre "chubascos atmosféricos extensos" a ser instalado en Bolivia.

Al inicio de las actividades del Laboratorio muchos científicos expresaron su deseo de visitar Chacaltaya y trabajar allí, otros fueron invitados por la Universidad. Ejemplos son la solicitud de Ted Bowen de la famosa Universidad de Chicago (en la que el profesor Arthur Compton, em 1941, decidió concentrar la mayor parte del Proyecto Manhattan, que atrajo a físicos de la talla de Enrico Fermi) hecha a Ismael Escobar en diciembre de 1952 y que resultó en su participación en los trabajos del Laboratorio, por un año, entre 1954 y 1955. También es interesante la nota del Rector Pedro Valdivia a Frank Oppenheimer, invitándolo a Bolivia.[51]

El Laboratorio y el Centro Brasileiro de Pesquisas Físicas

En el desarrollo de los acontecimientos que dieron origen al Laboratorio de Chacaltaya y su crecimiento, destacase la presencia del Centro Brasileiro de Pesquisas Físicas (CBPF) fundado en 1948 por un grupo de físicos brasileños, entre los cuales estaba Cesar Lattes, recientemente regresado de Bristol y los Estados Unidos.

"En el Laboratorio de Chacaltaya, montado en un esfuerzo de colaboración entre el CBPF y la Universidad de San Andrés, a una altura de 5 200 m en los

---

[50] Carta de F. Harris a I. Escobar de fecha 26 de agosto de 1954.

[51] La bien preservada correspondencia del Laboratorio que data de 1952 contiene una rica colección de cartas y notas de enorme interés histórico. Estos dos casos son apenas una muestra pequeña.

Andes bolivianos, preparan los Profs. Cesar Lattes, Ugo Camerini, Ismael Escobar, Alfredo Hendel, con un equipo de jóvenes investigadores brasileños y bolivianos, un programa de investigación que comprende entre otros, la determinación precisa de la vida media del mesón pi, por medio de circuitos de alto poder de discriminación, la medida de la densidad y del espectro de energía de los chubascos extensos y el estudio de las partículas inestables que acompañan los chubascos penetrantes, entre los cuales se encuentra el llamado mesón V y otros tipos de mesones menos conocidos; la determinación del segundo máximo de la curva de Rossi, etc. Los estudios sobre estas partículas inestables serán hechos con la Cámara de Wilson cedida por el Profesor Marcel Schein de la Universidad de Chicago; los otros estudios utilizan contadores y circuitos de alto poder discriminativo construidos en los laboratorios del Centro Brasileiro de Pesquisas Físicas en Rio de Janeiro." (cita del Prof. Costa Ribeiro en su artículo sobre la Física en Brasil en el libro Las Ciencias en Brasil; Fernando de Azevedo editor, 1954).[52]

Con los antecedentes de una estrecha colaboración entre Bolivia y Brasil, la Universidad Mayor de San Andrés encargó a una comisión compuesta por el Decano de la Facultad de Ciencias Sociales, Luis Ballivián Saracho, el Decano de Exactas Hugo Mansilla y el Director de Exactas, Jorge Muñoz Reyes, analizar un proyecto de convenio con el Centro Brasileiro de Pesquisas Físicas. Ante la opinión favorable a la Comisión, el Honorable Consejo Universitario de San Andrés aprueba la suscripción del convenio. En nota fechada el 3 de diciembre de 1952, el Rector Pedro Valdivia comunica a Cesar Lattes esta decisión. Inmediatamente después, Pedro Valdivia y Franklin Antezana Paz, Tesorero, por la Universidad y el Ministro João Alberto Lins de Barros, Presidente y Cesar Lattes, Diretor Científico, por el Centro, firman el Convenio de Colaboración.

Este Convenio permitió la obtención de recursos financieros y humanos de muy alta calidad y, lo que es más importante, permitió al Laboratorio convertirse en una facilidad científica de primera clase para el estudio de los rayos cósmicos. La cooperación científica con el Centro continua hoy después de 40 años haciéndola una de las más antiguas en América Latina. Sin embargo, no solamente el Convenio fue fundamental para el desarrollo del Laboratorio, sino también para el crecimiento del próprio CBPF.

Desde el punto de vista institucional, nota especial merece el apoyo que hasta principios de los años 70, prestó a la colaboración Bolivia – Brasil, el Correo Aéreo Nacional de la Fuerza Aerea Brasileña, transportando en sus vuelos mensuales, gratuitamente, personal y equipo. Los vuelos, al inicio en aviones DC-3, se iniciaban en el aeropuerto del Galeón en Rio de Janeiro, y concluian en La Paz.

---

[52] Marques A. "Dos Anos '50 aos Anos '90", Centro Brasileiro de Pesquisas Físicas, *Ciência e Sociedade*, CBPF-CS-0047/93, Rio de Janeiro 1993.

Al número de horas de vuelo se agregaban aquellas en que el avión permanecía parado en Corumbá o Sta. Cruz, dependiendo de las condiciones de tiempo de la cordillera.

Cuando se trataba de equipos más pesados se organizaba una auténtica expedición.[53] Ella incluía el tren Rio-Baurú, donde la carga era transferida a vagones del ferrocarril Brasil-Bolivia, siguiendo hasta Corumbá y Puerto Suarez; de Roboré a Pozo del Tigre se continuaba en camión para luego por tracción animal atravesar el Rio Grande. En la otra margen la carga se colocaba en jeeps o vagonetas que iniciaban su recorrido en el bosque conocido por Camino Real y terminaban en Santa Cruz, para luego por carretera seguir a Cochabamba y La Paz. Este es el camino que siguió por ejemplo, la cámara de Wilson cedida por Marcel Schein de Chicago a los primeros experimentos del Centro en Chacaltaya.

Varios nombres de científicos brasileños (y otros que trabajaban en Brasil) destacan como socios en la colaboración iniciada por Lattes en 1946 con su primer arribo a Chacaltaya. Algunos de los primeros, Giusepe Occhialini, Ugo Camerini y Roberto Salmeron, trabajaron estrechamente junto a catorce técnicos bolivianos. Ya se mencionó a Hervasio de Carvalho. En esta temprana época se montó una de las primeras Cámaras de Wilson, la que nunca funcionó a pesar de los esfuerzos desarrollados; ello no permitió explorar, como era el objetivo, una nueva frontera abierta por el descubrimiento de las partículas V por Rochester e Butler. Esta situación no comprometió sin embargo, otras iniciativas. G. Schwacheim y A. Wataghin, por ejemplo, desarrollaron un programa para estudiar la dependencia de chubascos penetrantes con la altura que fue pionero de los proyectos que más tarde se desarrollarían en El Alto y Chacaltaya. Una parte importante de todos estos esfuerzos fueron seguidos por la permanente presencia de Alfredo Marques.

Otro de los principales resultados de muchos años de colaboración entre la Universidad y el Centro fue ciertamente la puesta en marcha del proyecto denominado "Colaboración Brasil-Japón", al cual este ensayo se refiere en detalle más adelante. Tanto el Convenio con el Centro como la Colaboración Brasil-Japón contribuyeron enormemente a mantener estrechas relaciones entre las comunidades científicas de ambos países. Varios estudiantes bolivianos estuvieron en el CBPF y participaron de sus proyectos de investigación en rayos cósmicos. Algunos fueron: Magín Zubieta, Mario Bravo, Oscar Troncoso, Gustavo Pérez, Ricardo Anda y Rafael Vidaurre. Cesar Lattes fue claramente uno de los más decididos promoto-

---

[53] Una descripción completa de estas expediciones se encuentra en el Libro *Cesar Lattes 70 Anos: a Nova Física Brasileira*, A. Marques (editor), CBPF, Rio de Janeiro, diciembre 1994.

res de este intercambio.[54]

El crecimiento del Laboratorio

Entre 1952 y 1954, llegan a Bolivia expediciones de todas partes del mundo, para la ejecución de proyectos de investigación en rayos cósmicos y materias afines. Investigadores de Estados Unidos, Brasil, Gran Bretaña, Japón, India, Italia, Francia, Argentina, la Unión Soviética y muchos más conforman un impresionante grupo humano que, junto a sus colegas bolivianos, construyen la moderna ciencia de la radiación cósmica.

Cabe notar que el período 1952-1956, fue extremamente difícil por la situación social y económica ocasionada por la revolución del 9 de abril de 1952. En éste se tuvo que recurrir casi a la "caridad", como lo muestra la extensa correspondencia de la época. Ejemplos son, la solicitud de Escobar (denegada) a las fuerzas armadas para poder utilizar vehículo de las mismas durante los fines de semana, o la dramática nota enviada al Rector de la Universidad informando que los techos de Chacaltaya habían "volado" durante una tormenta de nieve y que era necesario adquirir calaminas para evitar tales accidentes. También, para el cableado de la red eléctrica del Laboratorio, se recurrió a material en desuso, entre este un transformador de la Prefectura del Departamento de La Paz.

A pesar de las dificultades, hacia 1954, la infraestructura del Laboratorio en Chacaltaya se había expandido; la oficina en La Paz contenía una importante biblioteca, salas de conferencias y otras. Ese año se completó la construcción de la línea eléctrica hasta Chacaltaya. Operó inicialmente a 6 kVA para luego pasar a 36 kVA. El material utilizado fue donado por el gobierno de Bolivia, de los restos de una mina de estaño. Personal del Laboratorio desmontó la línea de la mina y la transportó más de 160 km. La interconexión con la red local fue lograda por Juan Hersil, Alfonso Lazo y 8 trabajadores principales.[55] Fueron colocadas 226 torres y más de 2 000 kg de alambre de cobre.

El mismo año, el profesor Kurt Sitte de la Universidad de Siracusa, llega a Chacaltaya con otro experimento para estudiar los chubascos atmosféricos extensos. Al mismo tiempo, los Drs. Gottleb y Hertzler de la Universidad de Chicago montaron en Chacaltaya una de las primeras "Cámaras de Wilson".

---

[54] Es interesante notar la observación hecha por G. Bigazzi y J.C.H. Neto, que Lattes si bien es una autoridad celebrada en la física de partículas y la cosmofísica, fue también uno de los pioneros del método de datación por la observación del trazo de fisión. Él fundó dos grupos dedicados a esta disciplina, en Pisa, Italia, 1964, y en Campinas, Brasil, 1972 (*Cesar Lattes 70 Anos: a Nova Física Brasileira, op. cit.*).

[55] Fueron ellos: José y Eusebio Monasterios, Arturo Lopez, Tiburcio Quisucala, Vicente Cabaya, Marcos Calisaya, Honorio Paco y Ceferino Quilla.

Nuevos experimentos para medir el camino libre medio de la colisión en carbón y otros elementos fueron también instalados.

Los primeros resultados de las actividades del Laboratorio se publicaron en junio de 1952. Los trabajos (y monografías) resultantes de las actividades de Chacaltaya se han publicado sin interrupción desde entonces. Bajo el título de "Resumen de Labores", el Laboratorio presentó a la comunidad internacional sus principales trabajos. Esta revista científica es una de las más antiguas de la Universidad Mayor de San Andrés. El 17 de octubre de 1962, el Resumen recibió el Premio Nacional "Faja Verde", como la mejor producción científica y tecnológica de Bolivia. A pesar de que a mediados de los años 1970, la situación financiera en la Universidad impidió mantener regularidad en la publicación de Resumen, éste nunca perdió continuidad y hoy se hacen esfuerzos importantes para mantener su nivel científico.[56]

En colaboración con la UNESCO, en 1955, se organizó el Primer Curso Interamericano de Física Moderna. En él, físicos de Argentina, Bolivia, Brasil, Chile, Ecuador y Paraguay estuvieron presentes. Durante su estadía en La Paz, a propósito del Curso, uno de los instructores, el Dr. Mario Bunge de la Argentina, un distinguido filósofo de la ciencia, publicó su obra "La Edad del Universo".[57]

En 1957, también con la cooperación de la UNESCO, se desarrolló el Primer Simposio Interamericano de Rayos Cósmicos. En éste, cerca de cincuenta trabajos fueron presentados, no solamente por científicos de la región, sino de Inglaterra y de los Estados Unidos. Desde entonces la serie de Simposios Interamericanos tuvo siempre un carácter internacional.

Fue en este mismo año que el Physical Research Laboratory de Ahmedabad, India, envió a Chacaltaya un telescopio cúbico de mesones capaz de operar como un telescópio vertical de ángulo medio y cerrado. Este telescopio era similar en diseño a otros dos que ya trabajaban en la India y permitió una interesante comparación, respecto al fenómeno analizado, a diferentes alturas. El propulsor de este proyecto fue el distinguido físico hindú, Vikram Sarabhai quien lo discutió por primera vez con Ismael Escobar durante el Congreso Internacional de Rayos Cósmicos celebrado en México en 1956. El famoso físico Pierre Auger, entonces Director de Ciencias Naturales de la UNESCO, a quien Escobar y Sarabhai acudieron en pos de ayuda, facilitó los medios para hacer esta colaboración posible. En ella participaron a través de los años los Drs. Narayan, Nerurkar, Harjit Ahluwalia, M.S Danjhu y Lek Vir Sud.

---

[56] A.Velarde, "Instituto de Investigaciones Físicas", *Memorias de la Reunión de Editores de Publicaciones Científicas Andinas*, Universidad Andina Simón Bolivar y Comisión de las Comunidades Europeas, Sucre, noviembre de 1992.

[57] Bunge M., "La Edad del Universo", Publicación del Laboratorio de Física Cósmica, La Paz julio 1995.

Mis primeros trabajos en Chacaltaya, a partir de junio de 1962, los inicié precisamente con este grupo de la India. Durante los inviernos de los años 1962, 63 y 64, pasé varias horas tratando de mejorar el telescopio y entender la que para mi era entonces una muy compleja electrónica. En este mismo tiempo, se instaló un importante equipo para medir las variaciones temporales de la radiación cósmica, un monitor de neutrones tipo Simpson. Con él y con otro moderno que le sucedió, trabajaron durante años los físicos Ricardo Anda, Nicolás Martinic, Magín Zubieta,[58] Bruno Aparicio,[59] Eduardo Maldonado[60] y Oscar Troncoso.[61] También se instalaron telescopios direccionales Este–Oeste (Bruno Aparicio, Eliseo Guanca y Ricardo Anda).

A partir de 1956, no solamente el Laboratorio sirvió para observaciones de la radiación cósmica. Un equipo médico del Donner Laboratory, de la Universidad de California y los Drs. Herman Templin y William Talber del US Naval Ordinance Laboratory instalaron equipos para medir la componente infraroja de la radiación solar en altura. El US Naval Research Laboratory instaló equipos para medir la radioactividad atmosférica con los cuales se detectó la primera explosión nuclear francesa en el desierto del Sahara. Más tarde, las universidades de Manchester y Wisconsin realizaron experimentos en astronomía del infrarojo y otros temas conexos y el Instituto Boliviano de Biología de Altura, utilizó a los físicos y trabajadores de Chacaltaya para servir de "conejillos de indias" para sus primeros estudios sobre la adaptación y el comportamiento del hombre en la altura.

A partir de 1958, el Laboratorio incrementó sustantivamente sus actividades. La pequeña caseta en la cima de la montaña fue adecuada ya no sólo para servir como estación meteorológica, sino también para albergar nuevos equipos.

Entre la estación principal y la cima, en una pequeña mina abandonada, investigadores de la Universidad de Nuevo México instalaron un experimento para medir la interacción de los rayos cósmicos con elementos livianos y un telescopio para medir la variación de la intensidad de mesones respecto a la latitud. Estuvo involucrado en este experimento el Dr. Víctor H. Regener, colaborado por J. Kenny y Derek Swinson, y por los científicos bolivianos Abelardo Alarcón, José Antonio Zelaya y René Medrano.

Este mismo año, en El Alto de La Paz, cerca del Aeropuerto Internacional, George Clark del MIT y Juan Hersil, que acababa de regresar de los Estados Unidos, instalaron un importante experimento piloto para observar los chubascos atmosféricos extensos. En este experimento colaboró Gastón Mejía, quien se

---

[58] Este designado el 1 de septiembre de 1956 como Segundo Ayudante.

[59] Actualmente investigador en el Instituto Geofísico de Kiruna de Suecia.

[60] Quien continuaría sus estudios en el Imperial College for Science and Technology de la Gran Bretaña.

[61] Que pasó varios años en Brasil.

trasladó de El Alto a Cota Cota para participar en el experimento sobre ionósfera que se iniciara poco después, gracias a una colaboración del US Bureau of Standards, como parte de las actividades del Año Geofísico Internacional.

También fue asistente del experimento, y posteriormente el encargado del mismo a la muerte del Ing. Juan Hersil, el Ing. Ramón H. Schulczewski que fuera contratado para trabajar en un laboratorio de alto vacío que pretendía reacondicionar los tubos Geiger-Müller del experimento. Este fue un intento frustrado y Schulczewski se negó a cobrar sus primeros salarios.

Los éxitos alcanzados por el Laboratorio y aquellos que habían sido obtenidos durante años por el observatorio de San Calixto, dirigido primero por el Padre Descottes y luego, entre 1964-1993 por el Padre Ramón Cabré y junto a los esfuerzos del Instituto Geográfico Militar, constituyeron factores decisivos para el establecimiento y activa participación del Comité Boliviano en las actividades del Año Geofísico Internacional, uno de los esfuerzos cooperativos científicos a nivel mundial más importantes de las últimas cuatro décadas. El 15 de agosto de 1957, durante el gobierno del Dr. Hernán Siles Zuazo, se dictó el decreto supremo 04715 que crea el Comité. Mediante decreto 05725 el se transformó en Comisión Permanente que derivó institucionalmente en el Instituto Geofísico Boliviano. Esta década de actividades del Laboratorio concluye con un avance tecnológico de gran interés para Bolivia. La construcción del primer calentador solar plano por Mario Guzmán. Los rectores de esta época, Gastón Araoz (1954-1955) y, posteriormente, Ernesto Perez Ribero y Jorge Muñoz Reyes, juegan un papel importante de apoyo. Igualmente lo hace Angel Establier desde su cargo de Director de la Oficina de la UNESCO para Ciencia en Montevideo.

Figura 16.1: El inicio de la construcción de las edificaciones del Laboratorio.

Figura 16.2: La cúpula astronómica en la cima del Monte Chacaltaya.

Figura 16.3: Las edificaciones del Laboratorio.

Figura 16.4: Edificaciones del Laboratorio en los años 80.

17

CBPF: da Descoberta do Méson π aos Dez Primeiros Anos

Alfredo Marques

> ...*Anda a roda da História sem governo.*
>
> *As pegadas, o rastro, a referência,*
>
> *Quem os perdeu, perdeu a consciência.*[1]

Apresentação

O texto é uma tentativa de reapresentação dos primeiros dez anos do CBPF, ao ensejo dos cinquenta anos da descoberta do méson-π. Participaram dessa descoberta, além de *Cesar Lattes*, vários membros do laboratório *H.H. Wills* da Universidade de Bristol, dentre os quais seu Diretor, *Cecil F. Powell, G. Occhialini, H. Muirhead* e outros. Aqui se pretende relembrar o papel dessa descoberta na criação do CBPF e, através de sua atuação, na transformação do ambiente científico e universitário brasileiro da metade do século XX. Essas repercussões ficaram limitadas aos dez primeiros anos, mais precisamente, cobrem o período que vai desde a fundação do CBPF, em 1949, com o retorno definitivo de *Lattes* ao Brasil, até o ano de 1959, inclusive.

---

[1] Reynaldo Valinho Alvarez, *Galope do Tempo*, Tempo Brasileiro, 1997.

O texto está baseado principalmente nos relatórios do CBPF, publicados anualmente, cobrindo o período 1949-1959. Não foi possível localizar os relatórios referentes aos anos de 1950-52. Os dados sobre a fundação do CBPF e as projeções para 1950 aparecem no relatório de 1949; quanto aos demais, foram feitos de memória, por consulta a outros documentos e pelo relato de pessoas também presentes àquela época. Em princípio, as questões essenciais parecem não ter sido omitidas e foram registradas corretamente. Aqui se buscou principalmente recriar o clima sob o qual as atividades transcorreram, antes que o arrolamento dos personagens, estruturas e até mesmo datas, exceto quando disponíveis e necessárias ao propósito principal; o texto fica assim comprometido com a subjetividade das avaliações e seleções que tiveram de ser feitas para reproduzir esse clima. Uma descrição de parte dessas atividades, isto é, daquelas mais inseridas no meu horizonte de visão, fez parte de um documento anterior: Dos Anos '50 aos Anos '90, *Ciência e Sociedade 004/93*, em memória do Prof. *Francisco Mendes de Oliveira Castro.*

Dois grandes programas nacionais estavam em pauta no período coberto pelo texto: a Reforma Universitária e o Programa Nuclear Brasileiro, ainda que sem as formalidades que justificassem o uso desses títulos. Com o propósito de situar os desdobramentos da descoberta do méson-π no quadro mais amplo desses programas e de explicitar as relações com eles, dois capítulos foram incluídos: um sobre o estágio do desenvolvimento da universidade e da física brasileira à época daquela descoberta e outro sobre o quadro instalado no mundo com a libertação e controle da energia nuclear. Para estes as referências usadas foram, principalmente a) o texto de *Fernando de Azevedo* **A Cultura Brasileira**, cuja 6ª Edição acaba de ser publicada (1996), em coedição, pelas Editoras da UNB e da UFRJ; b) os depoimentos inseridos no texto do volume publicado pelo CBPF por ocasião da jubilação de *Lattes*: **Cesar Lattes-70 Anos: A Nova Física Brasileira**, *A. Marques*, editor, 1994; c) o artigo com que contribuí à LISHEP-1995: **Radiação Cósmica: Desenvolvimento e Interseções com a Física Brasileira**, *F. Caruso* e *A. Santoro* (eds.) no prelo;[2] d) o Livro de *Olympio Guilherme*, O **Brasil na Era Atômica**[3] Editorial Vitória, 1957; e) do livro de *Robert Jungk*, **Brighter than a Thousand Suns**, que tem como subtítulo A Personal History of Atomic Scientists, Pelican Books, 1965; f) do livro **Critical Assembly**, levantamento, a partir dos arquivos do Laboratório de Los Alamos, de todo o *Projeto Manhattan*, até a construção da Bomba Atômica, testes e lançamento, *Lillian Hoddeson, Paul W. Henriksen, Roger A. Meade, Catherine Westfall*, Cambridge University Press 2nd Ed. 1995. Outras referências aparecem nos rodapés.

---

[2] N.E.: Publicado em *G. Alves, F. Caruso, H. da Motta & A. Santoro* (eds.). *O mundo das partículas de ontem e de hoje.* 2ª ed. São Paulo: Livraria da Física, 2012, p. 11-44.

[3] Agradecemos a Cesar Lattes o acesso que proporcionou ao exemplar que possui dessa obra, hoje totalmente esgotada.

O texto está subdividido em capítulos, conforme o ano a que se refere; os dois primeiros capítulos, entretanto, cobrem períodos mais longos, relatando antecedentes da física brasileira e da física no mundo desenvolvido. Os limites colocados nas exposições desses capítulos foram dados por sua relevância para uma avaliação das repercussões da descoberta do méson-π no país e fora dele.

Física e Universidade Brasileira, dos Primórdios às Faculdades de Filosofia nos anos '50

• A melhor síntese do desenvolvimento da física brasileira, desde que apareceu como disciplina formativa nos cursos médicos das Escolas de Medicina do Rio e da Bahia, em 1832, é aquela dada por *Fernando de Azevedo* em **A Cultura Brasileira** (pg. 756, nota [20] ao capítulo "O Ensino Geral e os Ensinos Especiais"); ali relaciona Escolas, Institutos e numerosos professores que se destacaram no ensino da física mas a classifica, em todo o período, como "matéria de ensaio", não reconhecendo, até 1936, qualquer contribuição original à ciência física. Foi com a constituição do Departamento de Física da Faculdade de Filosofia Ciências e Letras da USP, através de *Gleb Wataghin, Marcello Damy* e *Mario Schönberg* que o país atingiu reconhecimento internacional nessa disciplina, modificando aquela contingência.

• A qualidade e atualidade dos trabalhos científicos realizados pelo grupo da USP, sob a direção de *Wataghin,* receberam as melhores referências durante o Simpósio sobre a Radiação Cósmica, realizado em 1941, com o patrocínio da Academia de Ciências, para marcar o fim dos trabalhos da missão chefiada por *Arthur Holy Compton*: lançamento de balões no interior de São Paulo para recolher dados sobre o efeito de latitude da radiação cósmica. Em seu discurso, *Compton* se refere aos trabalhos da USP de modo a não deixar dúvidas quanto ao seu valor científico e atualidade.

• Em sua análise, *Fernando de Azevedo* retrata o ensino superior brasileiro, o das ciências, em particular, como retórico, livresco, tendente à erudição bizantina. Mostra que, episodicamente se pode notar alguma excepcionalidade, dependendo de esforços e méritos pessoais de uns poucos, tudo retornando rapidamente ao padrão histórico, logo que esses personagens saem de cena. O preço pago pela cultura do refinamento literário sem qualquer reverência à importância do método científico, formou gerações de pessoas capazes de, segundo ele, "escrever sem pensar". Comparando a produção e qualidade científica de numerosos cientistas estrangeiros que passaram no Brasil parte de suas vidas com a de seus contemporâneos brasileiros, encontra-as desproporcionadas, qualitativa e quantitativamente, atribuindo-o à dominância dos fatores culturais enraizados no ensino.

• A propósito do Instituto Oswaldo Cruz assim se refere: "Se, no domínio das ciências aplicadas, já se havia desenvolvido, antes do advento das universidades, em 1934, as seções de pesquisa e experimentação em diversos institutos e se no Instituto de Manguinhos as atividades de pesquisa já haviam transformado essa escola de patologia experimental no maior centro de formação e de irradiação de cultura científica no país, quase tudo estava ainda por fazer nos diversos ramos da ciência pura, em que, com exceção das ciências naturais, sempre rarearam as contribuições originais dos brasileiros.

Nesse regime em que os problemas suscitados pela sociedade 'continuavam entregues aos práticos sem técnica e aos técnicos sem ciência', e em que dominava o interesse prático e utilitário do 'profissionalismo', cultivado através de mais de um século pelas escolas superiores de tipo profissional, não se compreendia facilmente que o estudo e o emprego das ciências aplicadas dependiam do conhecimento e do progresso das ciências puras".

• É notável o trecho de um discurso de *Ruy Barbosa* transcrito no capítulo sobre A Cultura Científica (nota [18]), pela contundência e perenidade: "Somos um povo de sofistas e retóricos, nutridos de palavras, vítimas de seu mentido prestígio, e não reparamos em que essa perversão, origem de todas as calamidades, é *obra de nossa educação*, na escola, na família, no colégio, nas faculdades. O nosso ensino reduz-se ao culto mecânico da frase; por ela nos advêm feitas e recebemos, inverificadas, as opiniões que adotamos; por ela desacostumamos a mente de toda ação própria; por ela entranhamos em nós o vezo de não discernir absolutamente a realidade, ou de não discerni-la senão através dessas 'nuvens' suscetíveis dos mais absurdos amálgamos e das configurações mais arbitrárias, em que a comédia de *Aristófanes* alegorizava a inanidade e as ilusões da escola dos sofistas no seu tempo". *Ruy Barbosa*.[4] Apesar das melhorias trazidas com a reforma *Chico Campos* que organizou as unidades isoladas existentes no Distrito Federal na primeira universidade brasileira, em 1931, do sucesso da criação da Faculdade de Filosofia Ciências e Letras da USP, em 1934, da sua congênere no Distrito Federal, em 1935, substituída em 1939 pela Faculdade Nacional de Filosofia, apesar de já se colocar em caminho programas de formação de professores para o ensino médio nessas unidades, segundo iniciativas de *Anysio Teixeira*, no Distrito Federal, e de *Fernando de Azevedo* em São Paulo e de já se colher os primeiros frutos dessas iniciativas, a situação, no início dos anos '50, na área da física, não era substancialmente diferente daquela no desabafo de *Ruy*.

• Talvez o relato mais eloquente da situação do ensino da física nessa época, decorrido já um tempo razoável após a instalação daquelas medidas, seja o de *Richard Feynman*, contido em seu livro de memórias.[5] Em suas visitas ao Brasil,

---

[4] Discurso pronunciado no Liceu de Artes e Ofícios em 23 de novembro de 1882. *In:* Orações do Apóstolo edição da Língua Portuguesa, Rio de Janeiro, 1923.

[5] Está a Brincar Sr. Feynman!, Ed. Gradiva, Lisboa 1988. Tradução de *Surely You're Joking Mr.*

em 1949 e 1951, 1953, falando fluentemente o português, teve oportunidade de dar aulas, conferências, participar de bancas examinadoras, para alunos da Faculdade Nacional de Filosofia e da Escola Nacional de Engenharia, inclusive em bancas do exame vestibular. Sintetiza suas conclusões no capítulo "O americano outra vez", onde começa narrando suas peripécias, tocando frigideira numa escola de samba (o título do capítulo é uma alusão ao desespero do mestre da bateria quando registrava a sua presença no ensaio, dada a frequência com que "atravessava" o ritmo em seu período de aprendizagem). Também narra sua experiência com estudantes. Seus comentários são muito semelhantes aos feitos por *Ruy Barbosa* setenta anos antes, mas foram tomados como destemperada impertinência. Suas críticas, sempre eloquentes, nunca foram levianas; resultaram de uma experiência mais ou menos prolongada e repetida com estudantes de diferentes classes, procedências e destinos. Não há dúvida que reencontrou, no começo dos anos '50, aquelas características culturais que vinham resistindo desde o século passado aos melhores tratamentos, quer pela reorganização universitária verificada a partir de 1930, quer pelos programas de formação de professores para o ensino médio levados a cabo nas Faculdades de Filosofia, o que é, de fato, chocante.

• Em meados dos anos '40, entretanto, já se notam no país, além do grupo da USP, pessoas interessadas no desenvolvimento da ciência e pequenos grupos científicos trabalhando apesar da atmosfera geral desfavorável; é assim que, no Rio, na recém-constituída FNFi, se encontrava *Joaquim da Costa Ribeiro*, autor de um trabalho científico da maior repercussão, a descoberta do efeito Termodielétrico, ou Efeito Costa Ribeiro, como ficou conhecido internacionalmente; no Instituto Nacional de Tecnologia *Bernhard Gross, Oliveira Castro* e outros que se notabilizaram com trabalhos de Reologia; o Núcleo de Matemática da Fundação Getulio Vargas que já na época publicava um periódico especializado, a *Summa Brasiliensis Mathematicae*, e mais *Luiz Freire*, na Escola de Engenharia, em Recife, garimpeiro de talentos científicos a quem o sul deve parte da melhor liderança na física e na matemática, *Francisco de Assis Magalhães Gomes*, em Belo Horizonte, e quem sabe mais quantos, espalhados nas longitudes brasileiras. Não estava claro, entretanto, face às dificuldades da época, que essas pessoas não representariam mais um surto histórico de qualidade, logo a seguir tragado nas fundas goelas do subdesenvolvimento histórico.

• Consciente das deficiências históricas, a *intelligentsia* brasileira amadurecia, ao longo dos anos '50, o aprofundamento das reformas de ensino colocadas pela revolução de 1930. Tratava-se, no limiar dos anos '50, de multiplicar e aprofundar os núcleos de pesquisas em ciência pura, em particular aqueles lidando com o método experimental, além dos limites já atingidos na Faculdade de Filosofia Ciências e Letras da USP e da FNFi, no Rio. A revitalização dos programas de

---

Feynman, compilação de conversas gravadas com *Richard Feynman* por *Ralph Leighton*, publicada em 1985.

formação de professores para o ensino médio foi também uma de suas metas mais importantes.

• Os anos '50 se caracterizaram por grande instabilidade política que se sucedeu ao suicídio de *Vargas* em 1954. O "golpismo" político, latente na sociedade brasileira, manifestou-se diversas vezes e sob diversas formas, tornando impraticável qualquer retomada de medidas educacionais de longo prazo que pudessem canalizar soluções para as demandas correntes. No decorrer do período setores organizados da sociedade enfeixaram todas as reformas, inclusive a universitária, num pacote amplo, conhecido como reformas de base. O modelo de universidade deixado por *Vargas*, burocrática, rígida e verticalmente hierarquizada passou a ser criticado por setores estudantis e boa parte do corpo docente que já reclamava medidas pedagógicas de modernização de currículos, laboratórios, etc. Com a decisão de *Juscelino Kubitschek* de transferir a capital federal do Rio de Janeiro para Brasília, a oportunidade de inovar apareceu no projeto de universidade para a nova capital, um modelo de universidade que incorporasse os reclamos de modernização. Essas ideias só começaram a se materializar na década seguinte mas foram atropeladas pelo movimento militar que depôs o Presidente *João Goulart*, em 1964. Vale mencionar que esse modelo de universidade tratava a ciência pura e a pesquisa, de modo geral, com grande prioridade e, dentro da ciência, a física, pelo que simbolizava de prestígio e dinamismo nas grandes transformações havidas no mundo, particularmente a partir da II Guerra mundial. Foi continuado no modelo da UNICAMP que se fundou sob diretrizes semelhantes, de prioridade à física e à ciência básica, mas que não resistiu às pressões históricas utilitárias, passando, anos depois, por transformações mais ou menos profundas que redirecionaram seu perfil original.

Projeto Manhattan

• A descoberta da radioatividade artificial pelo casal *Irène e Frédéric Joliot-Curie*, em 1934, pela absorção de partículas-α em Alumínio e outros elementos leves, levou *Fermi* e colaboradores, em Roma, a irradiar com nêutrons sais de Urânio e outros elementos pesados; a radioatividade obtida foi atribuída à formação de elementos transurânicos. *Idda Nodack*, ainda em 1934, mostrou em trabalho publicado no *Zeitschrift für Angewandte Chemie* que a radioquímica utilizada por *Fermi* era insatisfatória para justificar aquela conclusão, acrescentando que "quando núcleos pesados são bombardeados com nêutrons os núcleos em questão podem quebrar-se num número de partes grandes, sem dúvida isótopos de elementos conhecidos, mas não vizinhos dos núcleos submetidos à irradiação". Mme. *Noddack* anunciava assim a fissão nuclear, mas não foi levada a sério por *Fermi*, que tinha a seu favor a opinião da maior autoridade em radioquímica na época, *Otto Hahn*. O casal

Joliot-Curie que havia sido contestado por *Lise Meitner*, principal colaboradora de *Hahn*, no tratamento que os levou a concluir pela prova da radioatividade artificial, refez o experimento de *Fermi*, concluindo negativamente à formação de elementos transurânicos, dando início a uma disputa entre os dois grupos. *Otto Hahn* procurou dar fim a ela, optando por uma diminuição no ritmo dos trabalhos. Nessa altura, após a anexação da Áustria, *Meitner*, judia austríaca, teve de abandonar a Alemanha para fugir das perseguições nazistas, sendo substituída por *F. Strassman*. *Strassman* levou mais a sério os trabalhos do grupo de Paris, e insistiu com *Hahn* para a sua retomada, particularmente depois que *Irène* e *Savitch* anunciaram a presença de uma substância que se assemelhava ao Lantânio como subproduto da absorção dos nêutrons em Urânio. *Hahn* e *Strassman* retomaram as manipulações e demonstraram, com tratamento rigoroso, que a substância formada não era o Lantânio, mas sim o Bário; redigiram uma nota preliminar ao *Naturwissenschaften*, pouco antes do Natal de 1938, comunicaram o resultado a *Lise Meitner* que, com o sobrinho *Otto Frisch*, na ocasião fazendo-lhe uma visita em Estocolmo, interpretaram o fenômeno como o rompimento do núcleo de Urânio em dois núcleos menores. Deram ao processo o nome de *fissão nuclear*. *Frisch* comunicou esses resultados a *Niels Bohr*, então de saída para uma reunião científica em Washington; aí discutiu e deu ampla divulgação ao novo fenômeno.

• Possivelmente o único a acreditar em *Idda Nodack* foi *Leo Szilard*, refugiado húngaro, trabalhando em Oxford pois, ainda em 1934, apresentou uma solicitação de patente à Marinha Britânica para um dispositivo capaz de gerar enormes quantidades de energia, baseado na partição do núcleo de Urânio em seguida à absorção de nêutrons, embora não se descarte que tal posição, nitidamente visionária, fosse fruto de sua própria experiência. Logo a seguir, temendo a eclosão da guerra, emigrou para os EUA, instalando-se na Universidade de Columbia. A *Szilard* veio juntar-se *Fermi*, que em 1938, após receber o Prêmio Nobel, seguiu diretamente para os EUA, sem retornar à Itália; trabalhavam em laboratórios diferentes, no mesmo edifício e logo passaram a colaborar intimamente.

• *Fermi* e *Szilard* trabalharam intensamente no grande problema que se colocava na linha da liberação da energia da fissão: o de verificar se os produtos da fissão emitiam nêutrons em número suficiente com os quais se pudesse estabelecer uma *reação em cadeia*; obviamente, uma vez que mais de dois nêutrons fossem emitidos por fissão, na ausência de qualquer controle o dispositivo representaria um explosivo de colossal poder de destruição. *Fermi* e *Szilard* não foram os únicos a seguir essa linha de pesquisas; *Philip Abelson* no Lawrence Radiation Laboratory. *Norman Feather*, em Cambridge, membros da equipe do casal Joliot-Curie, bem como alguns pesquisadores na Alemanha e na União Soviética também trabalhavam intensamente nessas direções. Às vésperas de uma guerra de proporções mundiais, *Szilard* tentou convencer seus colegas de que os cientistas melhor fariam se mantivessem suas pesquisas em caráter sigiloso, abstendose de publicar qualquer resultado que pudesse servir a propósitos belicistas. Contou com o apoio de *Wigner, Teller, Weisskopf,* apoio que foi rapidamente se estendendo a outros físicos de origem européia e, embora com menor entusiasmo, a americanos, temerosos de que uma tal arma pudesse cair nas mãos de *Hitler.* Tentou inclusive contatar físicos alemães, encorajado pelas atitudes desassombradas de *Max Von Laue* contra o nazismo. Entretanto os esforços de *Szilard* não prosperaram: em abril de 1939, *Joliot, Hans von Halban* e *Leo Kowarsky,* da equipe *Joliot-Curie* encaminhavam à revista *Nature* o resultado de suas observações, dando como 3,5 o número médio de nêutrons emitidos por fissão (o número correto veio a ser posteriormente diminuído para 2,6). *Szilard* que também já tinha obtido esse resultado, foi levado a rever sua posição, substituindo-a pela de convencer autoridades militares americanas das potencialidades dessas descobertas. Na Alemanha os eventos tiveram curso diverso. Uma conferência secreta, convocada pelo Ministro da Educação, em 30 de abril de 1939, com a participação de seis físicos, para discutir o estado da arte nessa linha de pesquisas, no país e no exterior, teve seu conteúdo levado a *S. Flügge,* colaborador de *Otto Hahn,* muito conhecido como o Editor da prestigiosa coleção: *Handbuch der Physik. Flügge* teve reação oposta à de *Szilard*: considerou que o assunto deveria ter a mais ampla publicidade, para deixar de representar um perigo iminente e redigiu ele mesmo um artigo para o número de julho do *Naturwissenschaften,* dando uma detalhada descrição da reação em cadeia. O anúncio aumentou ainda mais o pânico na América; a construção de uma arma nuclear parecia então inevitável. Na Inglaterra as repercussões tiveram consequências mais práticas: *George P. Thomson* e *William L. Bragg* convenceram-se de que havia futuro na produção de energia a partir do Urânio e persuadiram o governo britânico a comprar e estocar mineral de Urânio, obtido principalmente da Bélgica, através de suas minas no Congo.

• *Fermi,* em seus estudos primeiros, observara que os nêutrons mais lentos eram mais eficientes que os mais rápidos na indução da radioatividade no Urânio. Até meados de 1939 nem o grupo de *Fermi* em Columbia nem o de *Joliot* em Paris, haviam obtido qualquer indicação de uma reação em cadeia, usando uma dispersão homogênea de Óxido de Urânio em água, que no caso servia para reduzir a velocidade dos nêutrons emitidos nas fissões (moderador). Foi quando *George Placzek* interpretou esse fracasso conjeturando que a quantidade de material fissionável e seu arranjo espacial no moderador não poderiam ser quaisquer: com demasiado Urânio os nêutrons seriam capturados pelo isótopo ^{238}U, presente em grandes quantidades no Urânio natural, antes de terem reduzido suficientemente sua velocidade; com pouco Urânio os nêutrons seriam capturados pelo Hidrogênio da água antes de produzirem novas fissões. Haveria assim, para cada arranjo experimental, uma *massa crítica.* A ideia de Placzek não foi bem recebida de imediato, pois faltavam na ocasião dados confiáveis sobre as seções de choque para captura de nêutrons em ^{238}U e em Hidrogênio; *Fermi* e *Joliot* continuaram seus experimentos com variações do dispositivo anterior mas nada conseguiram; *Szilard* tentou o grafite como redutor de velocidades mas não obteve amostras suficientemente puras e tampouco conseguiu diminuir as perdas de nêutrons no processo de redução da velocidade. *Fermi* resolveu abandonar essa linha de atividades e passou a se dedicar a problemas da radiação cósmica, aguardando que suficiente material fosse acumulado para um experimento em escala maior, num dispositivo heterogêneo, isto é, com blocos de material físsil concentrados em determinadas regiões do moderador; *Joliot* também logo desistiu do dispositivo homogêneo. Cerca de dois anos mais tarde, *Fermi*, com a colaboração de *Herbert L. Anderson* e *Walter Zinn*, construiu uma pilha heterogênea, com moderador de grafite e esferas de Óxido de Urânio (natural) e Urânio metálico (natural) da melhor pureza possível na ocasião, que veio a se tornar crítica, em 2 de dezembro de 1942; estava concluída com êxito a primeira reação em cadeia usando a fissão do Urânio.

• Justamente ao tempo em que o procedimento para a geração de uma reação em cadeia entrava no pior impasse, dias antes de França e Inglaterra declararem guerra à Alemanha, em setembro de 1939, *Bohr* e *Wheeler* publicaram sua análise teórica do mecanismo da fissão. Segundo esses estudos era o isótopo de massa 235 do Urânio que apresentava maior eficiência na fissão por nêutrons e não o de massa 238, muito mais abundante no Urânio natural. Esse resultado chamou atenção para a necessidade de enriquecer artificialmente as amostras de Urânio do isótopo ^{235}U. *Rudolf Peierls* e *Otto Frisch*, em Birmingham, utilizando dados disponíveis sobre seções de choque de fissão, calcularam a massa crítica (sem moderador) de amostras de Urânio natural e de Urânio-235 puro; a primeira resultou ser da ordem de toneladas, e a segunda de alguns quilogramas. Embora este cálculo fosse de interesse puramente acadêmico, pois ninguém imaginava tal nível de enriquecimento do Urânio, o resultado era encorajador para finalidades militares. *Peierls* e *Frisch* também estudaram teoricamente o processo de enriquecimento por difusão gasosa, concluindo que quantidades da ordem de quilogramas poderiam ser separadas em algumas semanas numa instalação com um número grande de seções de separação. Esses resultados juntamente com estimativas da operacionalidade de um artefato bélico foram levados a *George Thomson*, a quem o governo britânico fizera responsável por toda pesquisa envolvendo Urânio. Em 10 de abril de 1940 *Thomson* instalava o comitê britânico para o projeto de uma arma nuclear. Recebeu o codinome de projeto MAUD e contou com a presença dos físicos *Mark Oliphant, John D. Cockcroft* e *Philip B. Moon*; um subcomitê técnico foi instituído posteriormente, incluindo *Frisch* e *Peierls*. Um estudo para a operação de uma usina de separação isotópica foi iniciado na companhia Imperial Chemical Industries, por sugestão de *Frisch* e *Peierls*.

• Apesar da insistência de *Szilard*, a quem veio juntar-se *Fermi*, em sensibilizar setores governamentais para a importância dos trabalhos de pesquisa sobre a fissão do Urânio e seus desdobramentos, sobretudo militares, pouco obtiveram além de modestos subsídios para a continuação de seus trabalhos. *Szilard* sentia que a questão deveria ganhar a atenção das mais altas esferas governamentais, chegar ao Presidente *Roosevelt*, para receber o tratamento que demandava. Juntou-se a um conterrâneo seu, *Eugene Wigner*, radicado há mais tempo naquele país, e juntos concluíram que as autoridades mais altas dos EUA poderiam sensibilizar-se com uma manifestação de *Albert Einstein*, de longe o mais renomado cientista da comunidade. Após duas tentativas de chegar a *Roosevelt* usando diferentes intermediários (até o Cel. *Charles Lindbergh*, herói da aviação americana bem relacionado na colônia alemã, foi cogitado,[6]) fecharam, finalmente, na pessoa de *Alexander Sachs*, economista ligado a setores financeiros, que desfrutava da intimidade do Presidente. *Sachs* levou a *Roosevelt* um texto que, segundo *Szilard*, foi ditado por *Einstein* a *Edward Teller*, em alemão, passado para o inglês por *Szilard, Wigner* e *Teller*, assinado por *Einstein* em 14 de agosto de 1939.[7]

• Esse documento assinalava com muita ênfase o perigo de uma arma de colossal poder destrutivo ser desenvolvida na Alemanha, a partir das pesquisas sobre Urânio já em curso, dado que *Hitler* contava com grande acesso à comunidade científica, através de seu Subsecretário de Estado, pai do conhecido físico *Carl von Weiszäcker*, que se ocupava do assunto no Instituto Kaiser Wilhelm; pedia, em resumo, a concordância do Presidente para que o governo americano apoiasse essas pesquisas, dando-lhes a organização e os fundos suficientes para obter tal arma na América antes que os alemães o fizessem. Chamou também a atenção do Presidente para a necessidade de obter minério de Urânio, já que os alemães poderiam facilmente obtê-lo nas minas de Joachimstall, na Tchecoslováquia, então ocupada. *Roosevelt* contava com os assessores para assuntos científicos *Karl Compton* e *Vannevar Bush* que traziam a experiência desse cargo desde a 1ª Guerra, os quais, ouvidos a respeito, sugeriram ao Presidente que criasse o 'Comitê Briggs', encabeçado por *Lyman J. Briggs*, Diretor do Bureau of Standards, para dar consequência às colocações de *Einstein*. Esse Comitê foi posteriormente reorganizado, mercê, talvez, de uma segunda carta de *Einstein*, datada de 7 de março de 1940, e inserido na estrutura do National Defense Research Comittee. O *Projeto Manhattan*, estrutura que *Einstein* vislumbrou em suas cartas, capaz de abordar, com a organização e os recursos necessários, toda a gama dos problemas de pesquisas com Urânio, só foi, entretanto, constituído em 1942.

• Essas medidas foram consideradas apenas "mornas"; além da criação do Comitê e sua revitalização após a última carta de *Einstein*, as únicas medidas importantes se resumiram aos primeiros contatos para obter minério de Urânio e dar ao grupo da Universidade de Columbia, encabeçado por *Fermi* e *Szilard*, recursos para a compra de grafite e algum Urânio. Para alguns, como *J.R.Oppenheimer* e *Arthur Holy Compton* (irmão de *Karl Compton*) as cartas de *Einstein* não tiveram

---

[6] Ronald W. Clark, *Einstein, the Life and Times*, Avon Books, New York, 1972.

[7] *Ibid.*

maior importância fora da comunidade científica;[3] é que *Karl Compton, Vannevar Bush* e outros experimentados líderes em assuntos científicos e tecnológicos, que desfrutavam da intimidade de *Roosevelt,* mantinham contato estreito com os ingleses, na ocasião bem à frente dos EUA em tudo o que dizia respeito à produção de uma arma baseada na fissão do Urânio, mas que ainda não tinham respostas finais que justificassem os riscos e custos do empreendimento. *Roosevelt* e seus conselheiros se sensibilizaram mais profundamente apenas depois que uma missão britânica chefiada por *Henry Tizard,* Conselheiro Científico do Primeiro Ministro britânico, *Winston Churchill,* da qual fez parte *John D. Cockroft,* discutiu com membros do Comitê Briggs, em setembro de 1940, diante de uma pasta contendo seus últimos desenvolvimentos em armamento, em particular os baseados na fissão do Urânio, solicitando-lhes que se ocupassem desses projetos, de vez que os EUA tinham a capacidade industrial requerida e não estavam envolvidos na guerra. Vendo que a arma atômica não era apenas um delírio de *Leo Szilard,* que os ingleses estavam a ponto de demonstrar sua potencialidade e viabilidade, o Comitê Briggs decidiu intensificar o intercâmbio de informações com cientistas britânicos bem como apoiar outros grupos fazendo pesquisas com Urânio no país. Em meados de 1940 *McMillan* estudando o transurânico Netúnio-93, em Berkeley, observou uma atividade-α levando-o a crer ter descoberto o elemento de número 94; estes estudos preliminares foram continuados por *Seaborg, Kennedy, Segrè* e *Wahl* que, em fevereiro de 1941, anunciaram o novo elemento: Plutônio. De acordo com os resultados de *Bohr* e *Wheeler* o isótopo de massa 239 desse elemento seria também físsil, exibindo o atrativo de ser mais facilmente produzido por captura de nêutrons em ^{238}U e sucessivas desintegrações-β, dispensando os complicados, demorados e custosos métodos de enriquecimento do Urânio no isótopo ^{235}U. Ao mesmo tempo o projeto MAUD manifestava a palavra final de confiança na possibilidade de construção de uma arma baseada na fissão do ^{235}U, pois seu principal problema, a obtenção de quantidades ponderáveis de Urânio suficientemente enriquecido naquele isótopo, podia ser resolvido pelo processo da difusão gasosa. O esforço americano ganhou, a partir desse momento, novas dimensões, com a inclusão de novos grupos e universidades; representantes da indústria e líderes militares se juntaram para tratar das diferentes dimensões do problema, reorganizar e coordenar todo o trabalho existente. *Roosevelt* entregou a coordenação geral ao Gal. *George C. Marshall* que fez instalar, em junho de 1942, um grupo de engenheiros militares no bairro de Manhattan, em New York com esse propósito. O escritório ficou conhecido como "Distrito de Engenharia de Manhattan" e, em seguida, como "Projeto Manhattan". Foi dirigido pelo Gal. *Leslie Groves* a partir de 17 de setembro de 1942. É muito provável que informações auspiciosas de *Fermi,* demonstrando a obtenção de reações em cadeia evanescentes, porque abaixo da criticalidade (atingida somente em dezembro de 1942), tenham também contribuído para motivar *Roosevelt* à tomada de ações de maior alcance. Mas o principal agente deve mesmo ter sido a entrada dos EUA na guerra, em dezembro de 1941.

• O Projeto Manhattan foi o primeiro empreendimento humano de grande porte a lidar com o desconhecido; não teria sido possível, por isso mesmo, sem a presença de cientistas. Partiu da experiência prévia disponível nos EUA da prática da pesquisa no desenvolvimento de produtos em empresas de grande porte, como a ATT, GE, DuPont, e na experiência de gente da área acadêmica, como *Ernest O. Lawrence*, que na Universidade da Califórnia dirigia grupos de engenheiros e cientistas em projetos multidisciplinares, desde o de novos aceleradores de partículas até aplicações médicas. Custou cerca de 2,2 bilhões de dólares, ou seja, suas iniciativas contaram com recursos praticamente ilimitados, para a época, e incorporou 150 000 pessoas. Fortemente orientado para a missão em todos os seus setores, pesava sobre os ombros dos pesquisadores apenas o prazo para a finalização de sua meta principal, fixado por *Groves* em meados de 1945. Esse prazo foi cumprido quando a 15 de julho de 1945, no deserto de Alamogordo, o teste conhecido como "Trinity" foi realizado com sucesso, explodindo o primeiro artefato nuclear. Com a assinatura do armistício com o eixo Alemanha-Itália em 8 de maio de 1945, muitos cientistas ameaçaram deixar o projeto, de vez que sua motivação inicial já não contava. E muitos de fato o deixaram, dando início a uma crise que *Groves* e *Oppenheimer* contornaram com habilidade, limitando sua propagação. Ocorreram também manifestações diversas opondo-se a que se usasse a bomba para abreviar o conflito com o Japão. Para citar apenas duas das mais importantes, *Szilard* voltou a *Einstein*, agora para pedir-lhe uma carta a *Roosevelt* advertindo que uma explosão atômica sobre o Japão traria maiores problemas aos EUA do que o término da guerra por meios convencionais, mas *Roosevelt* faleceu a 12 de abril de 1945, sem possivelmente tomar ciência do conteúdo da carta. Em 11 de junho de 1945, um grupo de cientistas da universidade de Chicago redigiu e encaminhou ao Senador *Harry S. Truman*, que sucedeu a *Roosevelt* na Presidência, um longo documento[8] chamando a atenção para as imprevisíveis reações nos planos psicossocial e diplomático, com o previsível desencadeamento de uma corrida armamentista, caso a bomba-A fosse usada em território japonês. Aparentemente a decisão final já se tinha deslocado para o comando militar do projeto; entretanto, por recomendação do Secretário de Defesa, um comitê de alto nível, incluindo a presença de cientistas, conhecido como "Interim Committee", foi ouvido e se manifestou favoravelmente ao uso da nova arma em território japonês. Além do engenho testado no deserto de Alamogordo em 16 de julho de 1945, o projeto tinha mais dois artefatos praticamente prontos, pendentes apenas de arremates finais, um a ^{235}U, outro a ^{239}Pu, que explodiram em 6 e 9 de agosto, respectivamente, em Hiroxima e Nagasáqui. Essas explosões provocaram indignadas reações na comunidade científica internacional e em amplos setores da comunidade científica americana. Os desdobramentos futuros não foram tão brilhantes, porque marcados menos

---

[8] Conhecido como *"Relatório Franck"*, assinado pelo prêmio Nobel *James Franck*, e mais *D. Hughes, L. Szilard, E. Rabinovitch, T. Hogness, G. Seaborg, C.J. Nickson*, todos com participações de relevo no Projeto Manhattan. Era conhecido como o grupo de Metalurgia de Chicago.

pela engenhosidade e dedicação que por medidas de conteúdo policial necessárias para não deixar vazar o segredo da arma; as medidas foram fúteis, embora amargas para muitos, como previra o *Relatório Franck*, e desembocaram mesmo numa desenfreada corrida armamentista.

• O Projeto Manhattan foi um empreendimento de dimensões ciclópicas, marcando todo o desenvolvimento científico e tecnológico da segunda metade do século XX. Desde os problemas de purificação do Urânio a partir do minério, sua obtenção sob forma de compostos gasosos convenientes para os dispositivos de enriquecimento por difusão, até sua obtenção sob forma metálica, envolveram questões novas e delicadas soluções. O problema do enriquecimento só teve solução por uma combinação de separação de isótopos por métodos de difusão com o método eletromagnético, em separadores que usaram inicialmente os eletro-ímãs que *E.O. Lawrence* ia colocar num ciclotron,[9] alteração feita a meio curso de um subprojeto onde já se havia investido apreciável soma (a planta de difusão gasosa), o que demandou grande capacidade de diagnose, decisão e gerenciamento. O ^{239}Pu precisou ter *todas* as suas propriedades determinadas durante o projeto. A descoberta de sua fissão espontânea também gerou no projeto uma mudança de escala de grandes proporções, para desenvolver um detonador que aproximasse as partes reativas com velocidades muito maiores que no caso do Urânio, de modo a atingir a massa crítica antes que os nêutrons da fissão espontânea as consumissem. Foi "inventado" o primeiro reator nuclear e se produziu industrialmente água pesada e grafite de elevadíssima pureza; desenvolveram-se numerosos procedimentos químicos, metalúrgicos e físicos para enfrentar uma multidão de problemas novos. A determinação da massa crítica nunca teve um tratamento livre de questionamentos, quer porque não se conheciam suficientemente bem as seções de choque, especialmente para nêutrons rápidos, nem se poderia fazer experimentos para testar modelos teóricos sem que se arriscassem vidas. Assim, os métodos teóricos tinham que se basear em abordagens múltiplas usando o mínimo de generalidade, em benefício de resultados numéricos aproximados que ajudassem a dimensionar valores úteis; os métodos experimentais tiveram de ceder espaço dentro de sua lógica para a ingenuidade e improvisação mais típicas do inventor e, em muitos casos, buscar a segurança e confiabilidade características do engenheiro. Essas alterações na metodologia da pesquisa acadêmica que o projeto impôs, aliada à multidisciplinariedade dos grupos, à vontade de acertar, criou um clima de camaradagem e solidariedade entre os participantes do projeto que se procurou reproduzir em laboratórios nacionais do pós-guerra.[10] *R.R. Wilson*, tentou reviver na criação do FERMILAB um complexo científico reminiscente de Los Alamos: a metodologia empiricamente orientada, usada durante a II

---

[9] Chamou-se CALUTRON o separador eletromagnético, forma sintética de CALifornia University CycloTRON.

[10] Veja-se, a propósito, F.J. Dyson, *Perturbando o Universo*, p. 55, Ed. UnB, 1981.

Guerra, provou ser eficiente tanto pelo lado econômico como pelo lado do tempo despendido em Los Alamos e na construção do FERMILAB. A lição deixada pelo Projeto Manhattan é que ele foi muito mais que uma aventura de cientistas com explosivos: dotou os EUA de uma capacidade científica, de gerenciamento de grandes projetos, de produção de inovações tecnológicas em escala industrial, de novas máquinas e processos que influíram decisivamente em sua posição de nação hegemônica do Ocidente no pós-guerra. O grande beneficiário do Projeto Manhattan parece ter sido a empresa industrial. De fato, através dele aquilo que viveu apenas nos pensamentos de *Leo Szilard, Fermi, Otto Frisch, Rudolf Peierls*, e uns poucos outros, durante anos um mero sonho, foi transformado, no prazo estipulado, num "produto", deixando atrás milhares de soluções tecnológicas de alto valor, algumas das quais, complexas e gigantescas em si mesmas, autênticos sonhos, dentro do sonho. A grande empresa industrial passou a investir em pesquisa básica e, com sua capacidade de investimento e gerenciamento, mudou qualitativamente as relações com a universidade, passando de cliente a fornecedor de instrumentos, procedimentos e treinamento de recursos humanos. A receita "GG",[11] consistindo de fundos virtualmente ilimitados, prazos absolutamente improrrogáveis, projetos orientados pela missão dentro dos quais se associavam os esforços de cientistas, engenheiros, técnicos e administradores, pagou elevados rendimentos durante todo o período da guerra fria. Gradualmente os "scholars" foram sendo substituídos por cientistas com maior talento empresarial, capazes de gerenciar grandes grupos multidisciplinares.

1948: Ano da Gestação

• O méson-π, sua desintegração no mésotron e uma partícula neutra não identificada, sua massa medida com precisão estatística sobre um número de eventos enriquecido com exposições de emulsões nucleares em Chacaltaya, Bolívia, foi anunciado por *Lattes, Occhialini* e *Powell* em outubro de 1947, em trabalhos assinados pelos três, publicados na revista *Nature*.[12] Além de esclarecer todas as contradições envolvendo a componente dura da radiação cósmica (interação forte na produção múltipla, descoberta por *Wataghin, Pompeia* e *Souza Santos*, em São Paulo, em 1940, interação fraca na absorção nuclear, comprovada pelo experimento de *Conversi, Pancini* e *Piccione*, entre outras) teve ainda o atrativo de apresentar um valor da massa muito próximo àquele preconizado por *H. Yukawa* para o *quantum* das forças nucleares. Essa circunstância conferiu

---

[11] De "Greasy Groves", como era conhecido entre os cientistas.

[12] Um relato da descoberta do méson-π, muito autêntico e cheio de dados que usualmente escapam dos relatos científicos, foi publicado pelo CERN Courier, número de Junho de 1997, de autoria de Owen Locke, participante do grupo de Bristol à época da descoberta e portanto testemunha de todas as fases do evento. Reproduzido neste volume, pp. 23-26.

àquela descoberta enorme repercussão internacional; tratava-se presumivelmente de um passo colossal no entendimento das forças nucleares, exatamente aquelas "domesticadas" pelo Projeto Manhattan às custas de fabulosas somas de recursos de capital e humano, e de muitas vidas. O prestígio que o Projeto Manhattan conferira à ciência pura e seus praticantes não dava margem a vacilações; tratava-se, presumivelmente, da abertura de um novo caminho talvez mais fértil ainda e pleno de sedutoras surpresas que aquele aberto pela fissão do Urânio. Ninguém duvidava de que o intelecto dos físicos pudesse levar outra vez a fantásticas paragens, do maior significado para aprofundar o conhecimento da matéria até sua completa domesticação.

• Faltava, entretanto, uma confirmação importante: a de que essa partícula, até então limitada aos eventos revelados na natureza pelas interações da radiação cósmica com origem em outros recantos celestiais, acelerada e desviada para a Terra pelos caprichosos campos magnéticos dos espaços interestelares, pudesse também ser produzida numa máquina onde apenas a mão do homem exercesse todo o controle. Esse feito transcendeu à descoberta e foi realizado por *Eugene Gardner* e *Cesar Lattes*, usando partículas-α aceleradas a 380 MeV no ciclotron da Universidade da Califórnia, o mesmo cuja construção fora interrompida por *Ernest O. Lawrence* para, com suas peças polares, construir um dos calutrons, do Projeto Manhattan. A detecção do méson-π por *Gardner* e *Lattes* dependeu de muita engenhosidade e imaginação, para vencer as dificuldades relacionadas com o valor da energia de aceleração das α's, nominalmente abaixo do limiar para a produção de partículas com a massa necessária, e com a inexistência de feixe externo das partículas aceleradas. Essas dificuldades foram vencidas e as emulsões reveladas mostraram eventos onde foram produzidos mésons-π, tanto positivos como negativos. A repercussão desse feito foi ainda maior que a da descoberta do méson, pois colocava nas mãos da física um instrumento novo para a observação dos fenômenos subnucleares – o ciclotron. Tanto *Lattes* como *Lawrence* foram alvo das maiores atenções nos meios científicos, onde ganharam grande renome, convertido logo a seguir em popularidade nos EUA e fora. Nos EUA foram alvo de elogiosas citações e reportagens por parte da mídia não especializada. No Brasil *Lattes* conquistou grande popularidade, invulgar entre homens de ciência desde *Oswaldo Cruz*.

• A notoriedade fez de *Lattes* alvo preferencial para numerosas palestras e conferências, para estudantes, professores e para o público em geral, levando-o também, muitas vezes, ao consulado brasileiro em S. Francisco, onde desfrutou de agradável ambiente e estabeleceu muitos relacionamentos. Lá conheceu o poeta *Vinícius de Moraes* e o humorista *Millor Fernandes*, mas foi com *Nelson Lins de Barros*, funcionário do consulado como *Vinícius*, que suas relações de amizade tiveram maiores desdobramentos. Tiveram ocasião de conversar sobre os planos de fundar no Rio de Janeiro um instituto devotado às pesquisas em física moderna que pudesse abrir um espaço novo para a ciência e para a física em particular.[13] *Nelson* facilitou o encontro de *Lattes* com seu irmão Ministro *João Alberto Lins de Barros*, político de boa visão, com grande presença no desenvolvimento brasileiro desde os anos '30. Dadas as bem conhecidas repercussões da descoberta do méson, não foi difícil explicar a importância da iniciativa, bem como convencer *João Alberto* de que havia no Rio de Janeiro um grupo de jovens brasileiros competentes, cientificamente maduros, ansiosos por tal oportunidade. *João Alberto* garantiu trânsito à ideia em setores políticos da alta administração do país e, em 2 de maio de 1948 o CBPF começou a funcionar provisoriamente numa sala do 3° Andar do prédio n° 40 da Avenida Presidente Vargas. Em junho daquele ano o entusiasmo, o apoio recebido e o vulto de trabalho desenvolvido exigiram mais espaço, levando a sede das atividades para a Rua Álvaro Alvim n° 21 na Cinelândia, onde ficou até 1950, com o término do Pavilhão Mário de Almeida, construído em local cedido pela então Universidade do Brasil junto à sua Reitoria, no atual campus da Praia Vermelha. A organização da sociedade civil, diretoria, estatutos e demais organismos de direção só foram efetivados em 1949. A data de fundação do CBPF, aparecendo nos relatórios anuais, até 1954, é a de 4 de fevereiro de 1949; entretanto, a partir daquele ano, a data registrada é 15 de janeiro de 1949.

• O relato mais autêntico dos motivos e aspirações que levaram à fundação do CBPF, na palavra de seus responsáveis, foi registrado no Relatório Anual do CBPF, ano de 1949, transcrito abaixo:

Da necessidade de se fundar o C.B.P.F.

Como é do conhecimento público o Brasil já possui um grupo de pesquisadores com elevado nível científico, quer no campo da física, quer no da matemática.[14] Muitos desses cientistas tiveram oportunidade de se aperfeiçoar no exterior, graças a bolsas de estudos concedidas, sobretudo por organizações estrangeiras.

Em contato com os grandes centros científicos europeus e americanos, os físicos e matemáticos brasileiros ampliaram seus conhecimentos e, em bom número, destacaram-se de tal sorte que receberam propostas vantajosas na América e na Europa. Fato característico do entusiasmo que têm tido os bolsistas pelo progresso científico nacional é terem eles recusado essas oportunidades e voltado à Pátria, onde esperam aplicar e difundir os conhecimentos que adquiriram. Aqui, nossos cientistas, particularmente os que se dedicam à Física e à Matemática, encontram sérias dificuldades na falta de aparelhamento adequado, na inexistência de

---

[13] Essa era uma ideia acalentada por *Lattes* já antes da produção artificial do méson. Para um relato desses antecedentes ver o artigo de *J. Leite Lopes* em *Cesar Lattes 70 Anos: a Nova Física Brasileira*, Editor *A. Marques*, Rio de Janeiro, CBPF, 1994.

[14] Um depoimento significativo e valioso que testemunha o juízo elevado em que são tidos os Físicos nacionais encontra-se no número de julho de 1949 da revista americana "Reviews of Modern Physics" ("··· published to provide those interested in developments in physics with comprehensive and timely discussions of the problems which are of special interest and importance") Neste número, num artigo de *Hideki Yukawa*, físico japonês que no corrente ano foi contemplado com o prêmio Nobel de Física, aparecem citações de dez (10) trabalhos, publicados por quatro (4) físicos brasileiros.

bibliotecas especializadas e, o pior, na carência de regime de tempo integral que lhes permita dedicarem-se exclusivamente ao estudo e à pesquisa. Mas estes obstáculos, que são grandes, tornam-se pequenos em face do maior de todos, que inevitavelmente espera e abate todo o pesquisador de regresso ao Brasil: *a falta de ambiente isto é, a falta de compreensão e interesse que o meio demonstra para com o trabalho de pesquisa.*

Não desejando renunciar ao ideal que com tanto sacrifício abraçaram, os ex-bolsistas sentem-se impelidos, por sua perseverança, a continuar pesquisando em qualquer lugar e sob quaisquer condições. É claro que o pesquisador excepcional consegue produzir nas condições mais adversas, porém não há dúvida de que os resultados obtidos não correspondem à sua capacidade potencial. *Via de regra é o que tem acontecido entre nós, resultando num trabalho de baixo rendimento, de grande prejuízo para o País, que não aproveita eficientemente o saber e a capacidade de elementos tão valiosos.*

Para remediar esta situação precária, há muito um grupo de físicos e matemáticos brasileiros vinha almejando criar um centro de estudos e pesquisas destinado a proporcionar os meios materiais indispensáveis de que todos careciam. Mas só em fins de 1948, sentiu-se ter chegado a oportunidade de coordenar os esforços para conseguir-se a criação imediata da entidade há tanto tempo planejada.

O grupo de cientistas assim formado dirigiu-se às personalidades políticas, científicas, econômicas e sociais do País, destacadas pelo entusiasmo com que têm encarado iniciativas como essa. Mostraram-se elas de acordo com a premente necessidade de ser criado no Brasil um centro de pesquisas, e hipotecaram, desde logo, incondicional apoio a essa ideia.

Ficou assim assentada a fundação do C.B.P.F.

1949: Ano da Fundação

• O ano de 1949 foi de extenuante atividade que só a juventude e solidez de propósitos de seus fundadores permitiu realizar. Nesse ano, mais que o ideal de criar um instituto para conduzir projetos intelectuais de realização científica individual, lançavam-se as bases de uma organização capaz de lidar *dentro das condições brasileiras*, com aquelas reincidências históricas desfavoráveis do diagnóstico de *Fernando de Azevedo*: o CBPF sobreviveu a todas elas, participando ativamente da construção e consolidação da moderna pesquisa científica no Brasil. Não há dúvida, essa capacidade de sobreviver dentro da adversidade do meio brasileiro, revitalizada após breves períodos quando ganhava forças para voltar mais virulenta, deve muito ao espírito com que a instituição foi impregnada nesse primeiro ano de vida.

• Nesse ano se fez de tudo: seminários, aulas, pesquisa, busca de recursos para financiar o cotidiano, para uma nova sede – o Pavilhão Mário de Almeida – para a criação da infraestrutura, biblioteca, laboratórios de eletrônica, vácuo, oficina mecânica, de microscopia, para estudos com emulsões nucleares, Estatutos, câmara escura *etc.*, tudo com uma visão de prazo longo que transparece a cada linha daquele relatório. Em particular releva o cuidado com que se tratou da localização do CBPF, entre fundá-lo no Rio ou em São Paulo; é ainda o Relatório

1949 que faz o registro dessas medidas:

Da localização

Durante a fase inicial, estando em fase de organização e não possuindo ainda recursos suficientes para desenvolver uma vida autônoma, os locais mais propícios para a instalação do C.B.P.F. não poderiam deixar de ser a cidade de São Paulo ou o Distrito Federal.
De fato:
a) não possuindo patrimônio suficiente o C.B.P.F. não poderia começar pagando salários integrais a seus pesquisadores;
b) outrossim não tem o C.B.P.F. estabilidade bastante para oferecer posições seguras aos aludidos pesquisadores. É portanto indispensável que os mesmos estejam ligados a instituições que lhes ofereçam essa estabilidade e lhes permitam a tranquilidade de espírito imprescindível a quem se dedica ao estudo e à pesquisa;
c) uma instituição como o C.B.P.F. não pode manter vida isolada. Para obter os elementos humanos indispensáveis à sua vida e desenvolvimento o C.B.P.F., e o pessoal científico, técnico e didático do mesmo, devem estar em contato íntimo e permanente com uma entidade de ensino superior do tipo universitário, da qual sairão os futuros pesquisadores que irão integrar seus quadros.
Tendo em vista esses três pontos básicos, em março de 1949, a Diretoria do C.B.P.F., através de seu Presidente e de seu Diretor Científico, entrou em entendimentos com as Universidades do Brasil e de São Paulo, a fim de estudar as possibilidades de um trabalho de colaboração. O Magnífico Reitor da Universidade de São Paulo, Dr. *Lineu Prestes*, achou pouco oportuna a ideia de naquela Cidade instalar-se o C.B.P.F. em ligação com a Universidade, porquanto essa já possui Departamento de Física perfeitamente instalado e também por estar a mesma em condições de suprir todas as necessidades dos referidos departamentos. Na Universidade do Brasil, por outro lado, encontrou o C.B.P.F. um ambiente de grande receptividade para com seus objetivos, especialmente por parte do Departamento de Física da Faculdade Nacional de Filosofia, tendo o Magnífico Reitor, Dr. *Pedro Calmon*, dado desde o início seu irrestrito e mais entusiástico apoio, o que levou a Diretoria do C.B.P.F. a escolher o Distrito Federal como local para a instalação de sua sede.
Por sugestão do Prof. *Calmon*, o C.B.P.F. solicitou à Universidade do Brasil sua filiação àquela entidade sob o regime de Mandato Universitário, aprovado pelo Conselho Universitário em 7 de outubro de 1949.
Ficou do mesmo modo estabelecido que os professores e assistentes da Universidade do Brasil que desejassem realizar trabalhos de pesquisas no C.B.P.F. teriam livre acesso a seus laboratórios. Outrossim, os professores aceitos nos quadros do C.B.P.F. receberiam do mesmo um suplemento de salário igual a 1/3 (um terço) dos vencimentos de tempo parcial, comprometendo-se, em retribuição, a lecionar apenas uma cadeira na Universidade e realizar pesquisas apenas no C.B.P.F. ou num Laboratório da Universidade.

Assim, com Mandato Universitário da Universidade do Brasil, o CBPF mergulhava na questão da Reforma Universitária e, simultaneamente, no projeto nuclear brasileiro. Embora sua presença neste não tenha sido tão explícita, manifestava-se indiretamente no apoio recebido de amplos setores da sociedade mais sensíveis a ele, às promessas dele pendentes e aos esforços empreendidos pela maioria dos países para desenvolvê-lo desde o sucesso do Projeto Manhattan. Também conta para essa conclusão a polida recusa do Reitor da USP em acolher o CBPF; é que o Departamento de Física da USP já tomara medidas de reorganização e redirecionamento para o projeto nuclear, abandonando todo o sucesso científico obtido nos estudos da radiação cósmica em troca de uma liderança brasileira nas áreas nucleares, o que conseguiu efetivamente, o futuro o demonstrou. A tradição de trabalhos com a radiação cósmica sobreviveu, em troca, no CBPF, primeiro no laboratório de Chacaltaya que foi integrado ao CBPF como um de seus Departamentos, depois com os trabalhos de *Lattes* em voos de balões, em Minneapolis, com o Monitor a Nêutrons de *Georges Schwachheim*, e, finalmente, com a Colaboração Brasil-Japão.[15]

• A preocupação do CBPF com a manutenção e elevação do nível de qualidade de seus trabalhos propiciando a melhor formação possível para os candidatos à carreira científica, manifestou-se principalmente nesse ano de 1949, quando além de seus professores contou com a colaboração de renomados cientistas estrangeiros para a realização de cursos de especialização: *Cecille Morette*, do Institute for Advanced Studies de Princeton, deu um curso sobre Partículas Elementares; *Richard Feynman*, da Universidade de Cornell, deu um curso de Eletrodinâmica Quântica; *Francis D. Murnaghan* da Universidade John Hopkins, contratado pela Escola Técnica de Aeronáutica, deu um curso sobre Teoria das Matrizes; *Antonio Monteiro* deu um curso sobre Espaços de Hilbert e Desenvolvimento em Séries de Funções Ortogonais.

Uma mistura da competência com solidariedade também esteve presente nos primeiros anos do CBPF; cito apenas dois exemplos. O Relatório de 1949 registra 1 000 volumes na Biblioteca do CBPF, e acusa com prazer que, dentre eles, 200 foram doação de *Lourenço Borges*, Procurador do Instituto do Açúcar e do Álcool e redator do jornal "A Noite". Segundo: a seção de Microscopia, a cargo de *Elisa Frota Pessoa* e *Neusa Margem* (*Neusa Amato*, após o casamento), contava com material de primeira, chapas de emulsões nucleares expostas a prótons de 400 MeV, emulsões com mésons-π lentos, positivos e negativos, e outras com interações de nêutrons de 300 MeV, todas irradiadas no ciclotron da Universidade da Califórnia; tinha quatro microscópios para o trabalho de análise dos eventos, emprestados pelo Instituto de Química Agrícola e pelo Departamento de Polícia Técnica do governo federal.

• Outra questão da maior relevância que não deixou de sensibilizar o CBPF naquele primeiro ano de existência foi a de criar condições para a atração de estudantes, através da promoção de um programa de bolsas de estudos. Vale lembrar que não existia o CNPq nessa época; qualquer pretensão desse tipo teria de ser pleiteada junto a organizações estrangeiras. O CBPF instituiu um programa de

---

[15]A. Marques, Dos Anos '50 aos Anos '90, *Ciência & Sociedade* 004/93.

bolsas de estudos concedidas através de uma Comissão assim constituída: *Arthur Moses*, Presidente da Academia Brasileira de Ciências, *Bernardo Gross*, do Instituto Nacional de Tecnologia, *Carlos Chagas Filho*, do Instituto de Biofísica da UB, *Cesar Lattes*, do CBPF, *Ernani da Mota Rezende*, do Instituto de Eletrotécnica da UB, *Francisco Mendes de Oliveira Castro*, da Escola de Engenharia da UB, *Joaquim da Costa Ribeiro*, da FNFi, *José Leite Lopes*, do CBPF, *Lelio Itapoambira Gama*, do Observatório Nacional, *Leopoldo Nachbin*, do CBPF e da Escola de Engenharia da UB e *Luiz Cintra do Prado*, do Departamento de Matemática da USP. Era um programa tão útil quanto ambicioso:

...“Esse plano prevê a formação de pesquisadores por várias etapas. Partindo do estudante recém-formado pelo Curso Médio, o aludido Plano de Bolsas acompanha-o através do curso universitário, facilita-lhe os primeiros passos na pesquisa, segue-o durante as pesquisas de doutorado e mantém o pesquisador de valor sem preocupações financeiras até que o mesmo obtenha posição econômica condigna”...

A despeito dos poucos recursos e da criação, dois anos depois, do CNPq que assumiu explicitamente essas incumbências, o CBPF manteve essa iniciativa durante toda a década de '50. Ano de muita atividade de organização, de muitas projeções e planejamento, um dos pontos altos do Relatório de 1949 foi a proposta de atividades para 1950. Aí se encontra em detalhes a distribuição de verbas, de espaço e de responsabilidades com as novas unidades a funcionar em área de 600 m^2 no Pavilhão Mário de Almeida, com previsão para os primeiros meses de 1950. Um ponto dessa planificação merece destaque pelas repercussões que veio a ter, face a fatores não considerados mas que pesaram muito no desenvolvimento da iniciativa: o projeto de um ciclotron de 4 MeV para o CBPF. O limite de energia foi fixado por razões basicamente financeiras, segundo o Relatório Anual: “Seria de maior interesse um ciclotron para prótons de 40 MeV que, além de oferecer um campo mais vasto de pesquisas, é provavelmente o que permite a obtenção mais econômica de isótopos radioativos. Um ciclotron de 40 MeV custaria cerca de Cr$5.000.000,00, o que está fora de nossas possibilidades.”. Esse projeto foi cancelado anos depois, substituído por outro, sob a iniciativa do CNPq que se dispôs a financiar uma máquina de 20 MeV como ponto de partida para um acelerador maior, a ser construído assim que a máquina menor estivesse operando com físicos, engenheiros e técnicos treinados, que se pudessem ocupar de um projeto mais complexo.

• 1949 foi o ano da bomba-A soviética. A notícia explodiu com a potência de alguns quilotons no Pentágono e adjacências, embora não chegasse a constituir surpresa para os cientistas em Los Alamos. Essa notícia acabou sendo importante pelo que mostrou da inutilidade da tentativa de manter o segredo da arma nuclear trancado a quatro chaves, e, para o Brasil, porque aumentou o poder de barganha do país em áreas nucleares, pelo fato de ser um dos poucos detentores de reservas exploráveis de material físsil; foi um fator de aceleração das medidas que culminaram com a criação do CNPq, em 1951, e com ele a abertura de uma nova etapa para a condução dos assuntos da pesquisa científica brasileira.

1950: Ano do Mário de Almeida

• Ano marcado[16] pelos afazeres, despesas e canseiras da transferência do CBPF, das acanhadas dependências da Rua Álvaro Alvim 21, na Cinelândia, para o prédio construído em terreno da Reitoria da Universidade do Brasil, na Praia Vermelha. Com 600 m² conjugava modernidade nas linhas com a simplicidade dos materiais; alvenaria em poucas paredes, telhado industrial, muita madeira, compondo um ambiente onde transitar era fácil e a estada acolhedora. Foi construído com Cr$500.000,00 doados pelo industrial e banqueiro *Mário de Almeida* e recebeu seu nome, em agradecimento. Construído em dois pavimentos, o térreo passou a acomodar os laboratórios necessitando de maior desimpedimento para a entrada e saída de equipamentos, enquanto que o primeiro pavimento acomodou gabinetes de estudos, sala de reuniões, a Administração e a Biblioteca.

• No Plano de Trabalho e Orçamento para 1950, aparecendo no relatório de 1949, anuncia-se a criação de um Conselho Curador, "cujas finalidades precípuas seriam as de diligenciar e fiscalizar o emprego de capital social e opinar, obrigatoriamente, sobre os assuntos pertinentes à situação econômica e financeira do CBPF. A criação desse organismo foi motivada principalmente pela necessidade de criar uma estabilidade econômica maior para a instituição. Diz o referido Plano de Trabalho:

> A maior dificuldade encontrada na organização do programa de trabalho deve-se à inexistência de uma verba fixa anual que permita um plano de longo alcance. Sob este ponto de vista a situação do C.B.P.F. continua precária: basta lembrar que para o ano de 1951 não é possível fazer-se previsão orçamentária de espécie alguma....
> ...A campanha no sentido da obtenção de recursos para o CBPF tem sido feita, até o presente momento, pelos membros da Diretoria, ocasionando grave prejuízo sobretudo aos componentes dos quadros científicos, que se vêem obrigados a dedicar grande parte de seu tempo a esse setor, em lugar de dedicá-lo exclusivamente ao estudo e à pesquisa.

• O Plano de Trabalho e Orçamento para 1950 também menciona diversos cursos ministrados por pessoal do CBPF para alunos do curso de física da FNFi: Introdução à Mecânica, Introdução à Física Nuclear, Raios Cósmicos, Eletrônica, Eletrodinâmica Quântica, Métodos Matemáticos da Física e Alemão Científico, além de Colóquios e Seminários. Anuncia a compra de diversos equipamentos

---

[16] Os dados documentais usados para o ano de 1950 se restringem ao Boletim Informativo de no. 6, de junho daquele ano, e às projeções inseridas no Relatório de 1949 como plano de atividades para 1950. Foram completadas com informações obtidas pessoalmente.

de pesquisa e para as unidades de infraestrutura, bem como a aquisição de 5 000 volumes para a Biblioteca.

• Das atividades previstas para 1950 consta também uma capacitação em máquinas aceleradoras; o Boletim Informativo de junho de 1950 registra a viagem de um professor do CBPF à Holanda para tratar de diversas questões relacionadas à compra de um gerador tipo Cockcroft-Walton de 1,2 MeV à Philips e o Plano de Trabalho e Orçamento para 1950 consigna estudos para a construção de um ciclotron de 4 MeV. A construção dessa máquina foi negociada por *Lattes* com *E.O. Lawrence* e seus antecedentes constam da correspondência de *Lattes* com *Leite Lopes*, datada de janeiro de 1949 (Apêndice I). O Cockcroft-Walton em aquisição era de um tipo em final de linha para aquela categoria de máquina, mas muito útil para estudos de certas reações nucleares a baixas energias tipo (p, γ), cujo interesse científico atravessou a década de '50, além dos usos como fonte de nêutrons rápidos. Não havia como estender aquela máquina para energias mais elevadas, face aos problemas de isolamento elétrico que intervinham, além do que, nesse caso, não seriam competitivas com os aceleradores Van de Graaff, com melhores possibilidades nesse particular; a aquisição de um modelo comercial, assim, não frustrou planos de continuidade futura, dispensando consequentemente a preparação de técnicos capazes da construção de modelos com maior energia. Com o ciclotron a situação era diferente; face o sucesso da produção artificial do méson-π, a aceleração por máquinas circulares ganhou enorme impulso, particularmente nos EUA, na Inglaterra e na URSS; *Lawrence* e sua equipe já estavam lançando projetos com energias na região de alguns GeV, quando ofereceu a *Lattes* a oportunidade de treinar um pequeno grupo na construção de um acelerador de baixa energia, entre 2 e 4 MeV que, além do referido treinamento, seria útil para a produção de radio-isótopos; mencionou, conforme a correspondência citada, que três pessoas poderiam ter a máquina pronta em três meses, já que ele próprio acabara de completar, em seis meses, um modelo de 10 MeV também com uma pequena equipe. Tratava-se assim de uma proposta sumamente modesta, cuja finalidade principal era criar uma competência própria para investir em projetos futuros. Apesar da insistência de *Lattes* o plano para a construção dessa máquina acabou sendo cancelado e substituído por outro, com o apoio financeiro e direção executiva do CNPq. O cancelamento do projeto anterior se deveu a diversas causas, entre as quais a instabilidade do financiamento disponível orçamentariamente no CBPF, os empenhos e gastos com a mudança de sede, e a absorção das atenções de *Lawrence* por projeto na faixa dos GigaMeV, que já começara; a adoção de uma máquina com nova concepção e características dependeu da vontade do CNPq de começar com um projeto de grande impacto e das sugestões colhidas por participantes do Simpósio sobre Novas Técnicas da Física, havido no Rio, em agosto de 1952, particularmente dos físicos *I. Rabi*, de Columbia, *Herbert Anderson* e *John Marshall*, de Chicago, presentes ao encontro. O destino final dessas iniciativas não foi nada brilhante: o Cockcroft-Walton esperou anos por financiamento para um prédio adequado, já que seu terminal de alta-tensão exigia pé-direito e distância às paredes não convencionais. Terminou montado em dependência do IME, anos depois, quando sua utilização em física estava absolutamente superada, até mesmo como gerador de nêutrons, para o que já existiam comercialmente unidades mais compactas e eficientes. Teve alguma utilidade em testes de isolamento de componentes em alta tensão, sendo depois desmontado. Esse e outros fatos semelhantes, ocorridos em outras instituições, foram autênticas derrotas impostas pelo subdesenvolvimento. Além do Cockcroft-Walton do CBPF, o Betatron e o primeiro acelerador Van de Graaff da USP tiveram de aguardar períodos longos, da ordem de um ano ou mais, para que pequenos problemas de ordem técnica, como estabilização da frequência da rede ou o comportamento em clima úmido de materiais novos,[17] e ainda problemas de instalação adequada, pudessem ser gerenciados e resolvidos, problemas que nada tinham a ver com a atividade de pesquisa e, se não chegaram a comprometê-la, a atrasaram desmesuradamente; um Van de Graaff de 1,5 MeV da PUC/Rio esperou muitos anos pelo prédio para acomodá-lo e, quando finalmente montado, os pesquisadores tiveram que transformá-lo num acelerador para elétrons pois sua utilidade original já havia desaparecido totalmente. O projeto de ciclotron do CNPq colidiu, anos depois, com outra face do subdesenvolvimento brasileiro: o desvio de verbas. Explodindo em clima político altamente inflamado, que culminou com o suicídio do Presidente *Vargas*, o escândalo teve, além da paralisação dos planos do CBPF e do CNPq, o efeito de dividir e abalar a confiança mútua entre a liderança científica do CBPF, enquanto o CNPq mergulhava no descrédito para a condução do projeto nuclear brasileiro, em nome do qual justificara a entrada no programa de aceleradores. A divisão de opiniões na liderança do CBPF se deu entre os que defendiam que o assunto fosse tratado sigilosamente, enquanto se negociava a reposição dos recursos malversados, e os que não acreditavam ser possível evitar 'vazamentos" incontroláveis para a imprensa em clima político tão aquecido quanto o da época; além disso, as responsabilidades individuais pesavam desigualmente sobre os membros da liderança, e, portanto, os riscos assumidos eram, individualmente, também muito desiguais. O desfecho negativo acarretou o rompimento da unidade entre o projeto nuclear e o da reforma universitária que o CNPq incorporava; entretanto, a julgar pelas dificuldades econômicas e políticas que continuaram após a morte de *Vargas* e recrudesceram após o governo *Kubitschek*, é muito pouco provável que essa unidade pudesse ser mantida, fosse qual fosse o motivo de seu rompimento. A ideia, prevalecendo no CBPF, de que era necessário sustentar com autonomia pelo menos uma categoria instrumental no campo dos aceleradores sobreviveu e apareceu na década seguinte num programa bem sucedido de construção de aceleradores lineares, desta vez com o apoio da Comissão Nacional de Energia Nuclear.

---

[17] J. Goldemberg: "A Situação da Física Experimental no Brasil", *Ciência e Cultura* **12**, *n°* 1, 3 (1960).

• O CBPF assina com a Universidade do Brasil termo de acordo lavrando a outorga de Mandato Universitário. Segundo esse documento o CBPF se obriga a colaborar com a Universidade do Brasil, criando ou promovendo cursos cujos programas serão submetidos ao Conselho Universitário, franqueando laboratórios a professores, estudantes e técnicos credenciados pela Universidade, enquanto a Universidade obriga-se a reconhecer os cursos promovidos pelo CBPF. Foi firmado em 11 de agosto de 1950 pelo reitor em exercício, Prof. *Deolindo Couto* e pelo Diretor Científico do CBPF, Prof. *Cesar Lattes*. Assinam como testemunhas: *João Alberto Lins de Barros, Álvaro Alberto da Mota e Silva, Joaquim da Costa Ribeiro, José Leite Lopes, Gabriel Fialho, Nelson Lins de Barros, Ernani da Mota Rezende, Francisco Mendes de Oliveira Castro.*

• O Boletim n° 6 de junho de 1950 traz, entre outras, duas matérias que merecem comentários, por sua relevância. Uma é sobre a *Summa Brasiliensis Mathematicae*, dando um pequeno histórico daquela publicação:

> Em 1945 iniciou-se no Rio de Janeiro a publicação de uma revista matemática "Summa Brasiliensis Mathematicae", dirigida pelo Prof. *Lelio I. Gama* e publicada pelo Núcleo de Matemática da Fundação Getulio Vargas. Com a extinção desse núcleo, em 1946, Summa passou, em 1947, a ser publicada sob os auspícios do Instituto Brasileiro de Educação Ciência e Cultura, do Ministério de Relações Exteriores, comissão brasileira da UNESCO. A comissão de redação da Summa é constituída atualmente pelos professores Lelio I. Gama (diretor), *Antonio Monteiro, Francisco Oliveira Castro, José Leite Lopes* e *Leopoldo Nachbin* (secretário), estando o seu serviço de permuta e distribuição a cargo do Centro Brasileiro de Pesquisas Físicas. Toda correspondência relativa à Summa deve ser dirigida ao Centro Brasileiro de Pesquisas Físicas, Av. Pasteur 250, Rio de Janeiro.

Quer dizer, o grupo de matemáticos que encontrou na Fundação Getulio Vargas um precioso abrigo para o desenvolvimento de suas atividades para, assim, continuar aquelas tradições que começaram com *Joaquim Gomes de Souza* e estiveram sempre presentes na Escola de Engenharia, no Rio, perdeu todo o apoio em 1946, vindo encontrá-lo novamente no CBPF, onde estiveram sempre entre os mais ativos e onde fundaram o IMPA.

Segundo: esse número do Boletim traz o primeiro regimento da Biblioteca. A Biblioteca do CBPF não foi importante unicamente para seus pesquisadores. Foi um instrumento valiosíssimo para estudantes já que não havia nada de semelhante no Rio de Janeiro. Sua tradição de abertura para todos que a procuram vem dessa época e aparece em cada linha desse regimento.

1951: Ano do CNPq

• A criação do CNPq por ato do Presidente *Dutra* nos últimos dias de seu governo foi o fato científico dominante não apenas para o CBPF mas para toda a comunidade científica. Enquanto o CBPF, já na nova sede, dava continuidade a seus projetos, seus dirigentes suspiravam aliviados por passarem a contar com um respaldo governamental, através daquele órgão, que poderia suprir suas maiores carências para garantir o regime de tempo integral a seus pesquisadores e as perspectivas de uma ocupação estável. Foi seguida, dias depois, pela medida de criação da cadeira de Física Nuclear na FNFi, sendo nomeado para dirigi-la o Prof. *Cesar Lattes*. Nesse particular a descoberta do méson-π teve papel crucial. O representante brasileiro na Comissão de Energia Atômica das Nações Unidas, Alte. *Álvaro Alberto da Mota e Silva*, desde o ano de 1946 cogitava da constituição no Brasil de um órgão com aquelas características, isto é, de sistematizar, promover, fiscalizar, todas as atividades nas áreas nucleares. Esbarrava em vários obstáculos. Um deles, dos mais importantes, a dificuldade em convencer as autoridades políticas brasileiras de que o país já contava com uma massa de cientistas especializados capazes de dar início ao empreendimento. A despeito de se contar com *Costa Ribeiro, Wataghin* e seu grupo na USP, *Chagas* no Instituto de Biofísica, e outros, a reputação desses cientistas não ultrapassava os limites da comunidade, e, de outro lado, a grande parte do ensino superior da física estava nas mãos de médicos, muito competentes vários deles, mas sem maiores expressões fora do ensino tradicional, nenhuma, ou quase nenhuma, inserção em áreas nucleares que pudesse contribuir aos propósitos de *Álvaro Alberto*. A descoberta do méson, tendo *Lattes*, um brasileiro, entre os autores, e a enorme repercussão do feito no exterior deram a *Álvaro Alberto* os argumentos que lhe vinham faltando.

• Esse foi um ano de muita interação com o ensino no Departamento de Física da FNFi: *Jayme Tiomno* retornava dos EUA, após um período de estudos em Princeton, vindo juntar-se a *Leite Lopes* e, com a criação da cadeira de Física Nuclear, a *Lattes*, todos participando intensamente de programas de ensino, além de suas atividades de pesquisas no CBPF. *Lattes* tinha um ambicioso programa de pesquisas em Chacaltaya, pretendendo utilizar uma Câmara de Wilson que obtivera por doação de *Marcel Schein*, da Universidade de Chicago, em estudos das "partículas-*V*" descobertas por *Rochester* e *Butler*, operando uma câmara semelhante ao nível do mar. Deixou em seu lugar, parte do tempo, ocupado com as tarefas do curso, *Ugo Camerini*, contemporâneo seu no Departamento de Física da USP e no grupo de Bristol. Esses professores realizaram um trabalho de modernização do currículo sumamente importante para a formação de novos físicos. Em particular *Jayme Tiomno* e *Richard Feynman*, que voltara ao Brasil desta vez por mais tempo, em licença sabática, introduziram a ideia de livro-texto; por sugestão de *Feynman*, adotou-se o livro de *Slater & Frank* no curso de Eletromagnetismo, enquanto o velho compêndio clássico de *Eligio Perucca*, em dois volumes, era substituído por textos separados para Termodinâmica (*Zemansky*), Óptica (*Jenkins & White*), e outros; Física Atômica e Nuclear de *H. Semat* foi o texto adotado no curso de física nuclear e atômica. Essas modificações ajudaram a modernizar o ensino de física em todo o país, estendidas que foram a outras universidades graças ao continuado esforço daqueles professores. Um pouco mais tarde *Leite Lopes* deu um passo adiante: publicou um texto em português para o estudo de física atômica, recebido com grande euforia entre estudantes, muitos dos quais com dificuldades de realizar estudos em inglês.

• Ao tempo da criação do CNPq, pressões políticas se acumulavam em duas vertentes: a da reforma do ensino superior, com a necessidade de integrar a pesquisa ao ensino, particularmente em áreas de ciência básica, e a do projeto nuclear, sob o impacto das repercussões mundiais do Projeto Manhattan. O CNPq buscou incorporar as duas. A primeira linha dessas pressões já mereceu neste texto alguns comentários; o livro citado de *Fernando de Azevedo* dá muito mais. Quanto à segunda, vale a pena remontar às suas origens e discuti-la com alguma extensão.

• Quando a arma nuclear já se tornara uma realidade perfeitamente concebível, antes mesmo do teste Trinity, os EUA se preocupavam em mantê-la em segredo, impedindo sua reprodução em outros países. A forma usada para a manutenção desse monopólio alternou entre pressões ligadas a acordos internacionais, acordos militares bilaterais, onde sua posição vantajosa no domínio da tecnologia nuclear era barganhada com compromissos de fornecimento de matérias primas de valor estratégico e, finalmente, na legislação original da própria Comissão de Energia Atômica americana, condicionando a cessão de qualquer *know-how* nuclear a dispositivos contratuais que garantissem a ela a rigorosa observância da sua utilização. A manutenção do monopólio da arma atômica por um longo período era considerada inviável pela maioria dos cientistas; assim foi considerado no Relatório *Frank*, mas alguns setores militares não pensavam assim. Criada uma Comissão de Energia Atômica na Organização das Nações Unidas, passou esse órgão a receber propostas principalmente dos EUA e da URSS, visando ao estabelecimento de regras internacionais que levassem à não proliferação das armas, extinção dos arsenais existentes *etc*. Só que aos EUA interessava perpetuar a posição de vanguarda e à URSS igualá-la tão logo que possível, de modo que o confronto diplomático sobre as regras que deveria seguir a CEA das Nações Unidas passou a fazer parte do arsenal de disputa entre os dois países. O representante brasileiro, Almirante *Álvaro Alberto da Mota e Silva*, químico, inteirado das questões envolvendo o desenvolvimento nuclear, achava que o Brasil não deveria ficar alheio a esses desenvolvimentos; assim, defendeu acaloradamente a posição mais favorável aos países detentores de reservas de minerais físseis que almejavam ingressar de modo pacífico, mas soberano, no clube atômico. Essa posição o levou em certas ocasiões a votar contra propostas americanas, como quando estas procuraram defender o seu monopólio propondo a internacionalização das jazidas, ou seja, a perda do direito à propriedade para os países delas detentores. O uso de acordos militares para atingir esses mesmos fins partia do compromisso com premissas de solidariedade continental ou ideais culturais para estabelecer reciprocidade na troca de matéria-prima de valor estratégico, inclusive nuclear, por informação tecnológica que os EUA possuíam e cederiam na forma que lhes fosse mais conveniente, mantendo rigorosa fiscalização sobre seus usos de modo a garantir finalidades unicamente pacíficas. No caso das tecnologias nucleares, sua fiscalização para usos pacíficos impunha restrições tais que sua diferença de pura submissão ficava no nível semântico. Esse tipo de estratégia acabou muito abalado quando o Gen. *Dwight Eisenhower*, eleito presidente dos EUA para suceder a *Harry Truman*, denunciou o complexo industrial-militar, aliança de interesses industriais com interesses militares, visando a garantir lucros imensos aos setores que os dominavam; graças à reputação de *Eisenhower* conquistada durante a II Guerra, quando foi nada menos que o Comandante Supremo das forças aliadas, essa denúncia repercutiu em todos os países e deixou no limbo da dúvida a base moral daqueles tratados. O terceiro caminho envolveu a CEA americana, que estatizou todas as jazidas de materiais nucleares existentes, e por descobrir, em território americano, bem como técnicas e processos existentes, determinando ainda que a pesquisa de novos procedimentos só poderia ser feita por particulares com licença especial. A CEA também proibiu a cessão ou venda de tecnologia nuclear a não ser por meio de tratados ou convênios entre governos, nos quais ficasse garantido seu uso para fins estritamente pacíficos. Como, no caso nuclear, a linha que separa o uso pacífico da intenção bélica é imaginária, esse requisito exigia praticamente a submissão dos países, mesmo que desejassem unicamente aplicações pacíficas, a um tratamento discriminatório, incompatível com o desenvolvimento autônomo.

• A evolução dos acontecimentos nos anos seguintes mudou substancialmente esse quadro: a URSS detonou uma bomba-A em 1949, uma bomba-H em 1953, passando à frente dos EUA nesse artefato de guerra e colocou em órbita o primeiro satélite artificial da Terra em 1957. Neste ponto a "guerra-fria" internalizou-se, provocando extensa revisão da organização científico-tecnológica americana, inclusive intensificando a formação de PhD's: entre 1960 e 1970 os EUA fizeram mais PhD's que em todo o período 1880-1960! Perdido o monopólio da arma nuclear, muitas das restrições feitas anteriormente deixaram de fazer sentido; a não proliferação das armas foi atingida apenas recentemente, fora do plenário das Nações Unidas, que continuou palco do confronto ideológico ainda por muitos anos. Além disso, no caso brasileiro, pelo menos, o Acordo Militar com os EUA foi desfeito durante os governos militares, assim cancelando compromissos de fornecimento de materiais estratégicos. Foi criada a Agência Internacional de Energia Atômica em Viena, para substituir a Agência de Desenvolvimento Atômico da ONU, palco das maiores controvérsias durante o período do monopólio.

Naquela época, entretanto, as restrições eram de fato inaceitáveis pelos países que, detentores de reservas de materiais com valor nuclear, pretendiam desenvolver as técnicas para utilizá-los. Nesse particular, *Álvaro Alberto*, parece, percebeu com clareza a lição do Projeto Manhattan: mais que o desenvolvimento de uma arma de enorme poder destrutivo, ele propiciou o desenvolvimento tecnológico numa fronteira extensa e profunda, cujo significado revolucionou amplos setores da ciência e da produção industrial do pós-guerra. O Projeto Manhattan permitiu desenvolver, a partir da atitude em face do desconhecido, uma nova metodologia que revolucionou a tecnologia e a ciência em seus modos de produção. *Álvaro Alberto* utilizou amplamente, enquanto representante brasileiro na CEA da ONU, a figura das *compensações específicas*, segundo a qual não era o preço das matérias primas estratégicas, mas o que se poderia conseguir como *know-how* tecnológico em troca delas, o valor mais importante a ser buscado nas transações comerciais que as envolviam. O CNPq as incorporou desde o Estatuto original.

• Não tivesse sido a criação do CNPq o fato mais relevante do ano, este bem que poderia ficar conhecido como *Ano do Feynman*. Visitara o Brasil em 1949, por um período curto, participando de atividades didáticas e de pesquisas no CBPF, deixando-se contaminar pelo entusiasmo de seus colegas brasileiros no pioneiro empreendimento de formação de um instituto de pesquisas. Gostou e voltou, desta vez por mais tempo, em licença sabática. Aprendeu português rapidamente e, a não ser pelos hábitos rítmicos e pela acentuação que compunham um observável sotaque estrangeiro, falava fluentemente. Assim tinha condição de dar aulas, palestras *etc.* para alunos de graduação e para auditórios de não iniciados, mas interessados na física por qualquer motivo. Polemista inveterado, conversava com qualquer pessoa dentro e fora de sua área profissional; sempre risonho, muito simpático e afável, foi uma presença agradável e muito produtiva pelo tempo que despendeu no CBPF. Por seu jeito descontraído, amistoso, amante do diálogo, correram sobre ele muitas histórias; algumas foram registradas por *Ralph Leighton*, a partir de gravações a ele ditadas por *Feynman*, no livro onde apresenta as narrações mais deliciosas e extravagantes da vida do brilhante cientista (*Surely you're joking, Mr. Feynman*, publicado em português pela Gradiva, nota de rodapé n. 5). Dentre os *highlights* de sua vida registrados naquele livro, junto a referências à sua formação científica, afetos pessoais *etc.*, fez constar suas passagens pelo Brasil, o que não deixa dúvidas quanto ao apreço e estima pela terra e pela gente que conheceu. Foi um grande amigo do CBPF e do Rio, ou não teria retornado tantas vezes; em 1959, quando um incêndio destruiu livros e coleções de revistas da Biblioteca, atendeu prontamente ao pedido que lhe foi feito por *José Leite Lopes*, atuando junto a autoridades e colegas americanos até conseguir que a Fundação Ford doasse expressiva contribuição com a finalidade de repor as valiosas coleções destruídas.[18] Sua presença no CBPF ajudou muitos estudantes a construir uma

---

[18] J. Leite Lopes, "Richard Feynman in Brazil: Recollections", *Ciência & Sociedade* 03/88. Republicado no livro *Perfis*, F. Caruso & A. Troper (Eds.), Rio de Janeiro, CBPF, 1997.

atitude nova perante a ciência, valorizando o questionamento e a crítica antes da erudição.

• Para não encerrar este relato sem ao menos uma de suas numerosas histórias, passo a narrar uma que guardei na memória pelo inusitado de apresentar uma rara ocasião em que *Feynman* deixou alguém sem resposta. Ouvi-a de um de seus protagonistas, o Auxiliar de Portaria do CBPF *Fanor Rocha*. *Fanor* era desses tipos, conhecidos como "alegria de cartunista": quase se confundem com a própria caricatura. Ligeiramente estrábico e muito míope, via o mundo por trás de grossíssimas lentes que lhe impuseram o vício de erguer as sobrancelhas até o meio da testa para prestar atenção ao que quer que fosse; por isso exibia frequentemente um ar enigmático, mistura de soberba e espanto. Cabelos muito negros, lisos, arrumados prosaicamente "para trás", bigode espesso caído pelos cantos da boca completavam-lhe o semblante; anos depois, ganhava o apelido de *Janio Quadros*, tal a semelhança que guardava com o ex-presidente. Após uma reforma no interior do Pavilhão Mario de Almeida que terminou com a pintura esmerada de todas as paredes e portas (estas naquele tom rosa-tijolo da família imperial que o Reitor *Pedro Calmon*, historiador especialista nas coisas do 2° Império, exigia em todas as edificações da área), a Chefe dos Serviços de Portaria, D. *Maria Hercília*, pessoa muito rigorosa e severa, dera ordens expressas a *Fanor* de zelar pela conservação da pintura, qualquer violação ficando sob sua responsabilidade e ônus pessoal. Certo dia, terminada a aula no auditório situado no andar térreo do Mário de Almeida, mais ou menos a meio caminho no corredor que leva aos fundos, *Feynman* saiu em direção ao WC masculino, localizado no mesmo local que ocupa hoje. Como tivesse as mãos sujas de giz e a porta estivesse entreaberta, usou seu pé para empurrá-la, e assim entrar, sem deixar marcas brancas na porta. *Fanor* que tudo observava atentamente, de sua pequena mesa sob a escada em caracol que sobe para o primeiro piso, isto é, bem em frente ao delituoso feito, não teve dúvidas: partiu atrás do "infrator", interrompeu bruscamente seu hídrico alívio e, levando-o pelo braço até junto à porta declarou em tom solene, apontando para a maçaneta: – Moço isto aqui foi feito para abrir a porta com as mãos, não com os pés e, didaticamente, mostrou como usá-la. Virou as costas, saindo e retomando seu lugar de ofício, enquanto o rosto de *Feynman* varria todas as cores do visível, equilibrando-se finalmente no vermelho arroxeado. Quase apoplético deixou o prédio e pôs-se a andar pelas ruas internas do terreno da Reitoria, erguendo os pulsos crispados contra imaginários agressores, emitindo sons incompreensíveis que pareciam ameaças ou desafios contra os deuses, o que fez pelas seguintes duas horas, alternadamente com corridas e tentativas de chutes no traseiro dos gansos que o Magnífico Reitor mantinha na área para conferir-lhe ares de romana classe. No final de seu relato o bigode do *Fanor* recolheu-se discretamente para mostrar um sorriso de dever cumprido e triunfo caboclo.

1952: Ano do Simpósio sobre Novas Técnicas de Pesquisa em Física

• Realizadas suas tarefas iniciais de organização,[19] o CNPq logo a seguir mostrou a que veio: auspiciou a realização do Simpósio sobre Novas Técnicas de Pesquisa em Física entre 15 e 29 de julho de 1952, parte no Rio, parte em São Paulo. Compareceram ao Simpósio ilustres físicos brasileiros e estrangeiros, do mais alto nível: estiveram presentes *Sergio de Benedetti, J. Costa Ribeiro, David Bohm, Marcello Damy, Martin Deutsch, Bernhard Gross, D.W. Kerst, Cesar Lattes, José Leite Lopes, Marcos Moshinsky, Gert Molière, Giuseppe Occhialini, Isador Rabi, Emilio Segré, Jayme Tiomno, Manuel Sandoval Vallarta, Gleb Wataghin, Carl von Weiszäcker, Eugene Wigner,* entre outros. O encontro contou com 13 sessões, incluindo relatórios longos e comunicações curtas sobre o estado da arte em diferentes linhas como radiação cósmica, aceleradores de partículas, física de estado sólido, física nuclear e instrumentação. Esse encontro marcou também a interrupção do longo período transcorrido entre o último encontro internacional de físicos no Brasil, o Simpósio sobre Radiação Cósmica patrocinado pela Academia Brasileira de Ciências em agosto de 1941, que marcou o final dos trabalhos da missão *Compton.*

• Durante o Simpósio o CNPq tornou pública sua opção pela entrada na nova física, encomendando à Universidade de Chicago dois aceleradores, um de 21 MeV" que serviria de treinamento e outro semelhante ao de 170" construído em Chicago, alvo de um dos relatórios do Simpósio, a cargo de *Herbert L. Anderson. Anderson* passara à física de aceleradores, sendo o responsável pelo Sincrociclotron de Chicago, depois de trabalhar com *Fermi* no projeto Manhattan, participando da realização da primeira reação em cadeia auto-sustentada em dezembro de 1942. Junto com *Anderson* compareceu seu colega de Chicago, *John Marshall,* também da equipe do acelerador. A opção por essa linha de ação se deveu principalmente ao conselho do Prof. *I. Rabi* com quem *Álvaro Alberto* se avistara no ano anterior. A substituição dos planos originais de *Lattes* com a participação de *Lawrence,* em Berkeley, pode também ter tido propósitos de abrir novos caminhos para a recuperação das boas relações com a CEA americana, de vez que *Álvaro Alberto* adotara, nos estatutos do novo órgão, no que diz respeito aos materiais estratégicos de valor nuclear disponíveis no Brasil, a mesma atitude que a CEA tomara com respeito às novas tecnologias disponíveis nos EUA: estatizara-os. Além das grandes facilidades encontradas nas pessoas dos físicos americanos que trabalhariam em conjunto, houve, por exemplo, a concessão da Marinha americana cedendo

---

[19] Os documentos disponíveis para o ano de 1952 foram os Resumos do Simpósio, uma minuta manuscrita do Relatório do CNPq desse ano e o texto efetivamente publicado, e um documento do CBPF contendo apenas informações sobre pessoal. O documento sobre o Simpósio o devemos à cortesia da pesquisadora do MAST *Ana Maria Ribeiro de Andrade;* faz parte do acervo *Henry British Lins de Barros* daquele instituto.

gratuitamente todas as plantas do projeto do Sincrociclotron, cujo valor, nada desprezível, aproximava os US$500.000,00. Assim a política das *compensações específicas* de *Álvaro Alberto* dava um passo significativo, mesmo se tratando de uma área mais acadêmica, que não envolvia diretamente as aplicações pacíficas ou militares da energia nuclear.

• O folheto publicado pelo CBPF em 1952 não tem a forma dos relatórios anuais; além dos Estatutos, apresenta uma relação completa dos funcionários científicos, técnicos e administrativos, entre os quais se registram as presenças ilustres de *Richard Feynman, Giuseppe Occhialini* e *Gerhard Hepp*. O projeto de construção daqueles aceleradores dominava uma boa parte das atividades do CBPF, onde estavam instalados os laboratórios necessários e os técnicos ligados àqueles projetos, apesar de não ter implicado qualquer alteração das áreas de pesquisa e ensino que o CBPF vinha desenvolvendo desde a fundação. Foram criados os Serviços de Projeto e Construção do Sincrociclotron pelo Presidente do CNPq, com a incumbência de supervisionar e executar todas as tarefas de planejamento e construção do Sincrociclotron de 170", já que a máquina menor estava praticamente pronta,[20] devendo passar apenas por testes de ajustes e otimização de características e outras medidas visando exclusivamente ao treinamento de técnicos brasileiros. Nos relatórios consultados, o CNPq acena já com possíveis dificuldades: 1) a localização do acelerador pequeno, que se pretendia fosse no mesmo terreno do campus da Universidade, em área contígua ao Pavilhão Mário de Almeida, não teve a aprovação das autoridades universitárias. O relatório fala então em instalar esse acelerador em Niterói, em área a ser obtida do governo do Estado do Rio; este foi de fato o caminho seguido e, ano seguinte, uma parte do CBPF, oficina mecânica, eletrônica, laboratório de química, transferiu-se para Niterói, em terreno no morro de São João Batista vizinho ao Hospital. A localização do acelerador grande também foi avaliada, embora fosse problema mais distante, cogitando-se de terreno na Ilha das Cobras a ser cedido pela Marinha Brasileira; 2) o relatório menciona a disponibilidade de recursos de Cr$20.000.000,00 (cerca de US$1.000.000,00), mas menciona também que eles foram concedidos mas não efetivamente recebidos, todas as despesas até então correndo pelo Fundo Nacional de Pesquisas; 3) a questão da construção das peças polares para o acelerador de 170" foi discutida amplamente nesses documentos, do ponto de vista de fabricá-las no país, mas todas as soluções apresentadas assinalavam dificuldades impeditivas desse propósito; ainda com relação à fabricação no país de outros componentes o relatório menciona a dificuldade de obter, de modo completo, todas as componentes eletro-eletrônicas que o projeto consumiria. 4) O documento do CBPF de 1952 também assinala, sob forma de um aditamento, modificações em sua Diretoria para dar conta dos problemas de construção dessas máquinas: o Alte. *Álvaro Alberto*, Vice-Presidente,

---

[20] Depoimentos de pessoas envolvidas e outros relatos foram compilados por *Ana Maria Ribeiro de Andrade* no vídeo "Mésons, prótons, era uma vez um acelerador", MAST, 1995.

no exercício da Presidência do CBPF, designa, em 7 de junho de 1952 o Dr. *Álvaro Difini* para o cargo de Diretor Executivo do CBPF e, acumulativamente, de Diretor Tesoureiro, ocupados até então pelos Profs. *Hervasio Carvalho* e *Gabriel Fialho*, respectivamente. Embora essa medida visasse exclusivamente à agilização da administração geral dos recursos, colocando à sua frente pessoa experimentada, a circunstância de a mesma pessoa exercer também o cargo de Diretor Executivo dos Serviços de Construção do Sincrociclotron do CNPq esteve na origem de todo o abalo sofrido por esse empreendimento, pelo CBPF e pelo CNPq, associado ao escândalo do desvio de verbas.

• Dois importantes convênios foram assinados pelo CBPF com organizações de outros países em 1952. Um com a Universidade de Chicago, pelo qual o CBPF recebeu uma Câmara de Wilson a ela pertencente para instalação em Chacaltaya; outro com a Universidade Mayor de San Andrès, da qual faz parte o laboratório de física cósmica de Chacaltaya, por intermédio do qual o CBPF usava as instalações em conjunto com pessoal da universidade que participaria de todos os trabalhos em todas as fases. Mais tarde a presença do CBPF em Chacaltaya avançou para integrá-lo à sua estrutura departamental e assim facilitar as trocas de material, de pessoal técnico, pesquisadores e bolsistas. O chefe do Departamento foi o Prof. *Ismael Escobar* a quem *Lattes* conhecera desde a exposição de placas que consolidaram a descoberta do píon. Além dos trabalhos de instalação da câmara de Wilson em Chacaltaya, Escobar realizou trabalhos sobre chuveiros extensos, assimetria este-oeste, segundo máximo da curva de Rossi, entre outros. Dez anos mais tarde recebeu também as primeiras câmaras de emulsão, filmes de raios-X e chumbo dentro da colaboração Brasil-Japão sobre interações a altas energias na radiação cósmica.

• Este foi também o ano de criação do Instituto de Matemática Pura e Aplicada. Ocupando três salas do Pavilhão Mário de Almeida, congregava quase todo o grupo de Matemáticos do Departamento de Matemática do CBPF. Assim continuaram até que puderam dispor de uma sede alugada, na Rua Luiz de Camões, Centro do Rio.[21]

1953: Ano de Notas de Física

Com um trabalho de *Jayme Tiomno* e *Gabriel Fialho* o CBPF dava início à série *Notas de Física*. Destinada a divulgar os trabalhos realizados na instituição com maior presteza que as publicações convencionais, tinha o caráter de um *preprint*. A tiragem da revista, limitada inicialmente, chegou a atingir muitas centenas de

---

[21]N.E.: A referência não está correta. Na verdade, após a saída do CBPF, o IMPA transferiu sua sede para a Rua São Clemente e, só mais tarde, para a Rua Luiz de Camões. Dados mais completos sobre a História do IMPA se encontram em: L.A. Medeiros, *Aspectos da Matemática no Rio de Janeiro, Atas do XLVI Seminário Brasileiro de Análise*, Rio de Janeiro, novembro de 1997.

exemplares e, em meados dos anos '70, era alvo de resumos feitos por organizações destinadas à informação científica, no Japão e na URSS, pelo menos. A pré-condição para a publicação regular de Notas de Física foi a instalação de uma gráfica e o contrato de técnicos capazes de todo o trabalho de edição, datilografia de textos científicos, desenhos e fotografias, quando fosse o caso, revisão e reprodução. Esse trabalho foi feito sempre com um mínimo de funcionários muito dedicados e competentes. Pode-se dizer que foi um dos mais significativos feitos pioneiros do CBPF.

• Além de doze números de Notas de Física, a Divisão de Publicações alinhava a publicação do livro Introdução à Teoria Atômica da Matéria, de *J. Leite Lopes* e mais os seguintes, em fase de impressão: Física Nuclear Teórica de *R. P. Feynman*, Statistical Thermodynamics de *Leon Rosenfeld*, Fundamentos da Técnica de Vácuo de *Helmut Schwartz* e Curso Básico de Eletrônica de *Argus Moreira*. É importante notar o esforço corrente no CBPF para compor textos em língua portuguesa, para o qual o próprio *Feynman* contribuiu. Os textos em idioma estrangeiro, além de caros, recorriam a imagens ou referências nem sempre presentes na cultura de jovens graduandos, criando-lhes dificuldades maiores.[22]

Outro cuidado sempre presente foi com o Departamento de Ensino. Por ele passaram quase todos os membros da liderança científica do CBPF, *Lattes, Tiomno, Fialho, Camerini, Hervásio, Leite Lopes, Nachbin*. Em 1953 uma contratação veio dar novo estímulo à formação de um laboratório para práticas: *Paulo Emidio de Freitas Barbosa*, da cadeira de Física da Escola Nacional de Química, falecido recentemente, foi o nome escolhido. Sua atuação trouxe a marca da competência e do tratamento amigo, deixando recordações inesquecíveis em todos os que desfrutaram do treinamento ali proporcionado. Assumiu, anos depois, a cadeira de Física da Escola Nacional de Química, passando a outras mãos o Departamento de Ensino. Sua presença ficou marcada pela elaboração de um amplo elenco de práticas de laboratório, oferecidas aos estudantes de graduação. Essas práticas foram crescendo e diversificando, constituindo um conjunto muito completo, oferecido a um grande número de estudantes, já em meados da década. O Departamento de Ensino tinha também a seu encargo toda a administração de bolsas e demais contatos para estagiários em todos os níveis. Estagiaram, em 1953, 29 estudantes, entre alunos do curso de física da FNFi e outros da Universidade do Rio Grande do Sul (2), da Universidade de Minas Gerais (3), da Universidade do Recife (9), da Universidad Mayor de San Andrès (2), da Universidad de Buenos Aires (3), e da Boston University (1).

• Destacam-se entre os visitantes estrangeiros no CBPF em 1953 os seguintes nomes: *Richard Feynman* (CALTEC), *G. Molière* (Max Planck Institut für Physik,

---

[22] Um caso típico é a referência à forma do *doughnut* para figurar o toróide, quando o equivalente brasileiro dessa iguaria não tem nem de longe essa forma.

Götingen), *Leon Rosenfeld* (University of Manchester), *J. Robert Oppenheimer* (Institute for Advanced Studies, Princeton), *H. Joos* (Götingen), J.P. Davidson (EUA), *Leroy Schwartz* e *Richard Miller*, estes últimos técnicos em aceleradores circulares, ambos da Universidade de Chicago.

• O folheto NOTÍCIA publicado pelo CBPF em 1953 registra a seguinte lista de instituições com as quais mantém atividades em colaboração: Conselho Nacional de Pesquisas, Confederação Nacional da Indústria, Serviço Nacional de Malária, Instituto Oswaldo Cruz, Departamento Nacional de Produção Mineral, Escola Técnica do Exército, Diretoria de Estudos e Pesquisas Tecnológicas do Exército, Diretoria de Eletrônica da Marinha, Força Aérea Brasileira, Instituto Tecnológico de Aeronáutica, Universidade do Recife, Universidade do Rio Grande do Sul, Universidade de Minas Gerais, Universidade de São Paulo, Universidade do Brasil, Universidad Mayor de San Andrès (La Paz), University of Chicago, UNESCO.

• Essa publicação, além das atividades dos diferentes departamentos de pesquisas, do Departamento de Ensino, do Departamento Técnico, encarregado da instalação e construção dos ciclotrons do CNPq, da Biblioteca, do Departamento de Intercâmbio Cultural, refere-se à Divisão do Acelerador de Cascata, sob a chefia de *Peter K. Weyl,* que trabalhava na instalação do acelerador Cockcroft-Walton, adquirido à Philips e de um pequeno gerador de nêutrons.

1954: Ano de Mudanças

Em face das necessidades de espaço para a instalação do pequeno ciclotron não terem encontrado a concordância das autoridades universitárias, o CBPF transferiu parte de suas atividades para um terreno doado pelo governo do Estado do Rio de Janeiro em Niterói. Para lá se deslocaram as unidades de infraestrutura, o Departamento Técnico, com a Oficina Mecânica, o serviço de Eletricidade, e a Eletrônica, parte do Serviço de Alto-Vácuo, a Química, necessários ao funcionamento do ciclotron, e parte da Divisão de Raios Cósmicos que nessa ocasião trabalhava com instrumentação eletrônica para ser levada à Chacaltaya e dependia dos serviços de montagens eletrônicas. Vale a pena mencionar que a maior parte dos instrumentos usados na pesquisa tinham de ser feitos pelos próprios pesquisadores; os instrumentos comercialmente disponíveis limitando-se a contagens radioativas de rotina: nada de *scalers* rápidos, geradores de impulsos, amplificadores, coincidências rápidas ($\approx 10^{-9}$s). Mesmo fontes de alimentação em unidades separadas, capazes de alimentar um grupo de instrumentos não eram encontradas nas linhas comerciais de produção. A transferência para Niterói se fez com alto custo; anos depois os diferentes setores afetados ainda se queixavam dos recursos gastos com a transferência, retirados de seus magros orçamentos. A Divisão de Emulsões Nucleares, o Departamento de Física Teórica, o Departamento de Intercâmbio Cultural, com a Divisão de Publicações, o Departamento de Ensino e grande parte da Administração não se transferiram para Niterói, aguardando uma segunda etapa, após a entrada em funcionamento do acelerador.

• Essa etapa nunca chegou. Não fosse pelo abalo das relações entre o CBPF e o CNPq que se seguiu à descoberta do desvio de recursos dos Serviços de Projeto e Montagem do Sincrociclotron, envolvendo alto funcionário daquele órgão que exercia posição chave também no CBPF, teria sido pelo suicídio de Vargas, a 24 de agosto de 1954, que em curto prazo levaria o Alte. *Álvaro Alberto* a entregar o cargo de Presidente do CNPq em meio à completa descaracterização de sua política de compensações específicas por outras áreas da administração pública. Entre um episódio e outro aconteceu também o falecimento de *João Alberto Lins de Barros*, em 26 de janeiro de 1955, fundador e Presidente do CBPF desde a fundação, comoção de não menor importância, abalando a instituição em momento de tão grande fraqueza. O Relatório do Conselho Deliberativo referente ao período julho de 1954 a junho 1955, assinado por seu presidente, Prof. *Antonio José da Costa Nunes*, dá uma ideia do que foi a instabilidade institucional daqueles tempos:

> "No referido período, ocuparam o cargo de Diretor Executivo os seguintes titulares: Professor *Álvaro Difini*, Professor *Cesar Lattes,* Professor *Francisco Mendes de Oliveira Castro*, Professor *Carlos Chagas Filho* e Dr. *João Proença.* Com a renúncia do Dr. João Proença e vacância do cargo, o Presidente de então, Ministro *João Alberto Lins de Barros*, designou, nos termos do Art. 44 dos Estatutos o Dr. *José Machado de Faria para o cargo de Diretor Executivo desta entidade"*

• O mesmo relatório assinala, entretanto, alguns fatos muito auspiciosos: o primeiro deles foi a aprovação pela Câmara Federal de Deputados, a 1° de julho de 1954, da lei n° 2255, atribuindo ao CBPF as seguintes vantagens: 1) subvenção anual de Cr$10.000.000,00; 2) isenção de quaisquer impostos, direitos e taxas alfandegárias, exceto a de previdência social; 3) franquia postal e telegráfica; 4) licença de importação e cobertura cambial relativa a aparelhos, materiais, livros e publicações destinados exclusivamente às suas atividades científicas. O segundo foi a consignação no orçamento da Prefeitura do Distrito Federal, para o exercício de 1955, da importância de Cr$1.000.000,00, por iniciativa do então vereador *Paschoal Carlos Magno.*

• O relatório registra os nomes dos Deputados Federais que foram mais ativos na proposta e na apresentação de emendas à lei 2255 que a tornaram mais favorável ao CBPF. Dessa leitura não sobra a menor dúvida que o personagem-chave trabalhando por aquele desfecho fora o novo Diretor Executivo, *José Machado de Faria*. Exerceu o cargo de Diretor Executivo do CBPF desde sua nomeação, em 1954, até a sua incorporação ao CNPq, em 1976. Teve, assim, seu nome intimamente ligado às múltiplas estratégias de sobrevivência da instituição. Homem de perfil conservador, muito experiente na Administração Pública brasileira, com passagem pela Direção do DASP (Departamento Administrativo do Serviço Público, hoje ampliado no MARE), e pelo Departamento de Pessoal do Itamaraty, com grande mobilidade nesses setores e setores políticos correlatos, foi o administrador das carências, antes que dos recursos, do CBPF. Quem passa tanto tempo administrando carências não pode evitar de alinhar azedas críticas atrás de seus passos, frutos de numerosos interesses que precisou contrariar; junto a elas deixou, entretanto, muitos amigos que souberam compreendê-lo e apoiá-lo. O CBPF lhe deve a solução de diversas crises, mas sobretudo a visão de futuro e a ação diligente que o levou a garantir a escritura de posse do terreno da R. Xavier Sigaud, onde, em começos dos anos '60, fez construir o edifício ocupado até recentemente pelo LNCC e, à época da incorporação ao CNPq, o edifício Cesar Lattes, construído na sua estrutura básica. Na disputa por esse terreno *Machado* precisou enfrentar e vencer a cobiça de poderosos interesses, tendo nessa luta se empenhado pessoal e integralmente, jogando com toda a força que acumulou em sua área de influência durante longa vida pública. E, principalmente, em cerca de 22 anos à frente da Direção Executiva do CBPF que assumira na esteira de um rumoroso caso de desvio de verbas, jamais se soube de qualquer acusação ou mesmo dúvida sobre sua lisura na aplicação dos recursos da instituição.[23]

• O Relatório desse atormentado período assinala, entretanto, atividades de pesquisa e ensino aparentemente indiferentes às amarguras e turbulências à sua volta. O Departamento de Física Teórica, por exemplo, menciona uma atividade excepcionalmente elevada nesse período, a julgar pelo número de publicações científicas. Os Departamentos de Raios Cósmicos, Química e Radioquímica reclamaram da falta de recursos, da separação dos laboratórios e de algumas precariedades em Niterói, embora mantendo seus padrões de trabalho. Essa foi, aliás, uma característica essencial no núcleo da capacidade de resistência à adversidade que o CBPF demonstrou naqueles anos; mesmo quando alguns setores ficaram fortemente abalados por terem sido mais atingidos diretamente, como o foram aqueles que tiveram de enfrentar uma ida-e-volta a Niterói em pouco mais de um ano, todos mantiveram o padrão e, em alguns casos, até intensificaram a atividade.

• Finalmente, aquele relatório assinala com números frios um elenco de pequenas doações quase simbólicas, mas extremamente expressivas pelo toque de autenticidade e postura solidária: subvenções aprovadas para o CBPF pelas câmaras de vereadores de pequenos municípios, ainda hoje com muitas carências,

---

[23] Na verdade houve um caso de denúncia anônima, que à época dos governos militares era possível; *Machado* foi ao Ministro da Educação, o Cel. *Jarbas Passarinho*, para lhe solicitar pessoalmente a abertura de inquérito com vistas à apuração total das denúncias. Acompanhei-o nessa audiência e testemunhei seu empenho para que o inquérito fosse instaurado e levado às últimas consequências. O inquérito saiu e nada apurou que desabonasse sua conduta.

no orçamento das respectivas prefeituras, testemunhos do elevado conceito gozado pela instituição, de sua penetração e do quanto se julgava importante o apoio ao desenvolvimento científico. Honraram o CBPF com suas atenções os seguintes órgãos municipais: Prefeitura Municipal de Itaboraí, Câmara Municipal de Presidente Wenceslau, Câmara Municipal de Jardim, Câmara Municipal de Barra do Piraí, Câmara Municipal de Gurupi, Prefeitura Municipal de Igarapé Mirim, Câmara Municipal de Angra dos Reis.

• 1954 foi o ano da 1ª Edição de As Ciências no Brasil, coletânea de artigos aos cuidados de renomados cientistas sobre as ciências, seu desenvolvimento histórico, suas instituições e suas práticas no Brasil, até aquela data. O Editor, *Fernando de Azevedo*, escrevera anteriormente A Cultura no Brasil, já referido e, quando o revia para uma atualização, resolveu destacar tudo o que se referisse às ciências num volume separado; não se sentiu capaz de redigi-lo, convidando especialistas de cada área para fazê-lo. O texto sobre A Matemática no Brasil é de autoria de *Francisco Mendes de Oliveira Castro*, Professor Titular do CBPF, falecido recentemente. A Física no Brasil coube a *Joaquim da Costa Ribeiro*, da FNFi, onde se encontram diversas referências aos trabalhos realizados no CBPF até a data.

1955: Ano da Conferência de Genebra

• Por iniciativa da Organização das Nações Unidas, teve lugar em Genebra uma Conferência Internacional sobre Usos Pacíficos da Energia Atômica. A conferência teve grande repercussão e um amplo atendimento por parte da grande maioria dos países membros da ONU. O Brasil mandou numerosa comitiva. O CBPF foi representado por *José Leite Lopes*. O Relatório Anual se refere ao seu licenciamento entre 1 de maio de 1955 a 31 de agosto do mesmo ano:

> quando exerceu o cargo de Secretário Científico da ONU para a Conferência Internacional sobre as Aplicações Pacíficas da Energia Atômica.

Publicou, a respeito da Conferência, um longo artigo em Ciência & Cultura. A Conferência analisou, entre outros temas, o esgotamento das reservas de combustíveis fósseis do ponto de vista geo-econômico, prevendo uma crise energética dentro das primeiras décadas do século XXI, quando o uso da energia nuclear se tornaria imperativo. Por ora a participação da energia de origem nuclear no suprimento das necessidades mundiais está em torno de 17%, com nítidas pressões para aumentar.

• A Conferência de Genebra marcou o lançamento de planos de desenvolvimento das aplicações pacíficas da energia nuclear em todo o mundo; logo a seguir se criava a EURATOM e a Agência Internacional de Energia Atômica, que veio a substituir o organismo de controle criado na Comissão de Energia Atômica da ONU; também logo a seguir se criava o CERN. A própria CEA americana recebeu emendas de modo a reformar seu estatuto original, estatizante e marcadamente monopolista, de modo a acomodá-la às novas realidades. Além desses fatos auspiciosos que representavam o portal da era nuclear, a Conferência recebeu muitas manifestações de protesto de cientistas com relação ao armamento nuclear, desde as explosões de Hiroxima e Nagasáqui até os testes para o desenvolvimento de novos artefatos que estavam poluindo descontroladamente o planeta. Foi entregue ao Presidente da Conferência, Prof. *Homi Bhabha* uma carta aos cientistas de autoria de *Bertrand Russell* subscrita por *Einstein*, pouco antes de falecer, condenando a guerra e solicitando ao plenário que assim se manifestasse junto aos governos dos países ali representados.

• A turbulência institucional no CBPF continuou em 1955; o Presidente do CBPF, Ministro *João Alberto Lins de Barros*, após viagem ao exterior para tratamento da saúde, reassume seu cargo a 20 de dezembro de 1954 e falece a 26 de janeiro de 1955. O Vice-Presidente, Alte. *Álvaro Alberto da Mota e Silva* declina de assumir a Presidência pelo fato de já se ter exonerado da Presidência do CNPq, em 13 de janeiro, e não achar conveniente assumir a presidência do CBPF. Uma eleição é então feita da qual resulta sufragado o nome do Prof. *Elisiário Távora*, para terminar o mandato de *João Alberto*. A 18 de agosto de 1955 o Presidente do Conselho Deliberativo, Prof. *A. J. da Costa Nunes* encaminha o relatório anual daquele Conselho ao Presidente do CBPF, Gen. *Edmundo de Macedo Soares e Silva*, eleito no mês de junho para substituir o Prof. *Távora*. Assim, com duas eleições e dois Presidentes em cascata no intervalo de seis meses, encerrava-se o período de maior turbulência. Mas as feridas abertas levaram muitos anos para cicatrizar.

• Um fato auspicioso e importante ocorrido esse ano foi a criação pelo CBPF do Museu de Ciência, em 27 de outubro de 1955. A Prefeitura do Distrito Federal apoiou a medida e indicou uma Comissão, que incluía membros do CBPF, para sugerir medidas para o funcionamento do Museu dentro do menor prazo possível (24/01/56). O Museu nunca saiu do papel mas é interessante reproduzir aqui o pensamento da casa quanto às finalidades desse órgão:

> O desenvolvimento futuro das atividades do Centro dependerá, entre outros fatores, da quantidade de cientistas, engenheiros e técnicos nacionais disponíveis para seus trabalhos. É este justamente o problema de mais difícil solução porque depende de fatores imponderáveis, estranhos mesmo às disponibilidades financeiras Edificações e equipamentos podem ser providenciados em curto prazo desde que, para tal, haja verba bastante e suficiente. Um pessoal selecionado constitui, todavia, empreendimento a longo prazo e que depende, também, do interesse que os jovens tenham pela ciência e tecnologia. O Centro vem, desde há muito, se preocupando com esse problema, porque vê, com apreensão, que o número de jovens ora interessados nos seus ramos de trabalho não será suficiente para qualquer expansão em suas atividades.
> Assim, como um auxílio à solução desse problema, e como empreendimento

para recrutamento desse pessoal, resolveu o Centro, em meados do ano p.p. criar o Museu de Ciência. O Centro espera que o Museu transforme a natural curiosidade das crianças e dos jovens em interesse científico, proporcionando assim, futuramente, um maior número de cientistas, engenheiros e técnicos especializados. O Museu propiciará também às instituições de ensino, de qualquer grau e tipo, o uso de seus laboratórios, seus equipamentos, seus filmes, etc., auxiliando assim o ensino dos jovens, o que, indiretamente fará aumentar o número de interessados em ciência e tecnologia.

Outras medidas serão necessárias para resolver o problema dos futuros cientistas, engenheiros e técnicos, mas é fora de dúvida que a principal delas é a do perfeito e eficiente funcionamento das instalações e serviços do Museu de Ciência...

• O relatório 1955/1956 assinala nove cursos programados para o período 1 de setembro de 1955 a 31 de julho de 1956 dados na FNFi e na Escola Nacional de Engenharia. O plano de atividades para o período seguinte, entretanto, se refere a dificuldades com os cursos dados na FNFi, originadas na falta de um convênio específico com aquela unidade, em complementação ao mandato universitário, que tornasse oficiais as disciplinas cursadas e efetivos os créditos dos alunos aprovados. Um convênio desse tipo já fora firmado com a Escola de Engenharia. Essa medida foi de fato tomada, em 1956, e o problema solucionado. Assinou-se, em seguida, um convênio semelhante com a Escola Nacional de Química.

• O Prof. *Cesar Lattes* licenciou-se para passar um período em Minneapolis.

1956: Ano da CNEN

• O biênio 1954-1955 fora extremamente difícil para o CNPq à frente do projeto nuclear brasileiro. A implementação da linha das compensações específicas com que o Alte. *Álvaro Alberto* pretendeu nortear a venda de minérios de valor nuclear aos EUA encontrou forte oposição naquele país e poucos aliados no nosso. Perdido o monopólio da arma, os EUA continuaram obstinadamente negociando a compra de materiais sensíveis para aumentar seus estoques, em vista do surto previsível das aplicações pacíficas e para dificultar o desenvolvimento de armas nucleares por outros países além da URSS. Como as reservas desses minerais não eram muitas e poucos os países detentores, grandes pressões foram exercidas para obter vendas desses materiais sempre, é claro, nas condições mais vantajosas. As "compensações específicas" do CNPq colidiam com artigos da lei McMahon que criou a CEA americana e também com compromissos bilaterais assumidos em face de tratados, como o Battle Act, de 1951, o Acordo de Assistência Militar Brasil-EUA, de 1952 e outros decorrentes de reuniões de ministros plenipotenciários dos dois países, Notas Diplomáticas, enfim, um cipoal jurídico que deixou pouquíssimo espaço para as conquistas visando à autonomia tecnológica antes do esgotamento dos recursos em minerais de uso nuclear. O que se conseguiu foi realmente muito pouco: além dos planos frustrados do sincrociclotron de 170", a compra de um reator de pesquisas, enquanto que do outro lado houve troca de monazita por excedentes de trigo, sem qualquer compensação fora do plano da retórica diplomática. Em 1953 *Álvaro Alberto* resolveu recorrer a outros interlocutores. Contratou com uma firma francesa a produção de Urânio metálico de alta pureza. Em 1954, contratou na Alemanha o fornecimento de um jogo de ultracentrífugas com as quais poderia enriquecer Urânio no isótopo U-235, enriquecimento esse que, embora em nível baixo ($\approx$1%), seria suficiente para alimentar um reator a água leve para produzir energia elétrica – sugestão que lhe fora feita pelo Prof. *J.R. Oppenheimer* quando visitou o Brasil, em 1952, a convite do CNPq. As dificuldades para *Álvaro Alberto* aumentaram a partir dessa encomenda. Em novembro de 1954, o Gen. *Juarez Távora*, Chefe da Casa Militar do governo *Café Filho*, que assumiu a presidência após a morte de *Vargas*, recebe, ao que consta, de seu primo, Prof. *Elisiário Távora*, quatro documentos procedentes de meios diplomáticos americanos, dois datados e assinados, dois sem data e apócrifos, veiculando as críticas mais veementes às negociações com o governo brasileiro para o fornecimento de materiais de valor nuclear e, em particular, colocando o Alte. *Álvaro Alberto* como principal empecilho aos entendimentos; o último dos documentos era um veto à compra das ultracentrífugas (então já pagas) alegando que o *status* da Alemanha em face do Tratado de Armistício assinado em 1945 não a permitiria transferir tecnologia nuclear a outros países sem licença americana, licença que jamais seria dada.[24] Juntamente com o impacto desses documentos, encaminhados ao Conselho de Segurança Nacional para a elaboração de uma nova política nuclear, o país reincidia na turbulência política, com a queda de *João Café Filho* e sua substituição, primeiro por *Carlos Luz*, depois por *Nereu Ramos*, que finalmente marcou as eleições dos quais saiu vitorioso *Juscelino Kubitschek*, em outubro de 1955. Antes disso, em 13 de janeiro de 1955, *Álvaro Alberto* pedia exoneração da Presidência do CNPq, tendo ficado até março do mesmo ano, quando foi substituído pelo Sr. *Batista Pereira*.

• Empossado a 31 de janeiro de 1956, já em abril daquele ano, JK criava uma Comissão especial para estudar e propor nova política nuclear; revogava o Acordo de Prospeção Conjunta de Urânio e em 10 de outubro desse ano criava a Comissão Nacional de Energia Nuclear, com responsabilidades normativas e executivas para todos os assuntos que se referissem à energia nuclear, incorporando muitas diretrizes do Estatuto do CNPq, no que se aplicava; em 31 de agosto desse ano, era criado o Instituto Nacional de Energia Nuclear, hoje IPEN. O CNPq ficou, a partir daí despojado de sua ramificação no projeto nuclear brasileiro. Concentrou-se no projeto de reforma universitária, mais antigo porém com definições menos

---

[24]Esses documentos foram tornados públicos após denúncia do Deputado *Renato Archer* ao plenário da Câmara Federal, em agosto de 1956. Mais de trinta anos depois, o grupo da Marinha que trabalha no projeto do submarino nuclear anunciava a operacionalidade de um sistema de separação de isótopos por ultracentrifugação. Projeto e construção próprios.

claras, mergulhado em todo o complexo da educação no país. Sua participação foi, entretanto, decisiva para a inserção da pesquisa científica no ensino de 3º grau, particularmente a pesquisa em ciência básica; aqui a física e o CBPF tiveram destacado papel.

• Segundo a brasilianista *Nancy Stepan*, o segredo para o sucesso do Instituto Oswaldo Cruz, manifestado pela reputação alcançada por aquele instituto, dentro e fora do país, foi ter aliado a um projeto de pesquisa em ciência básica, um projeto de ciência aplicada, com projeções sociais e econômicas bem visíveis. Sugere em sua análise que essa deva ser uma fórmula geral a ser perseguida por outras instituições brasileiras no caminho de seu reconhecimento. Acatada a validade dessa análise, a separação do projeto nuclear do projeto de reforma universitária foi um fato negativo. Entretanto, a julgar pelas dificuldades econômicas e políticas que se instalaram no país, sobretudo após o governo JK, é muito difícil imaginar que esses projetos tivessem podido caminhar juntos. A ciência básica na universidade brasileira sempre teve, entretanto, em seu interior, um projeto de grande alcance econômico e social: o da formação de professores para o ensino médio, colocado nos anos '30 com toda a ênfase por *Anysio Teixeira*. Durante os anos '50 esse projeto ainda foi considerado como parte dos compromissos da comunidade universitária das áreas científicas básicas; na reformulação dos anos '70 esse compromisso perdeu muito da nitidez, ao mesmo tempo que a categoria profissional de professor de ensino médio despencava aos níveis mais baixos de organização e remuneração.

• O Departamento de Física Teórica contou com a visita de *Gleb Wataghin*, entre outubro e dezembro de 1956. Além de encantar a todos com sua presença e personalidade, participou ativamente de todas as atividades do Departamento, tendo ainda publicado um trabalho de pesquisa.

• O Departamento de Ensino registra oito cursos de graduação dados por professores do CBPF para alunos de física da FNFi e do curso de Engenharia Nuclear da Escola Nacional de Engenharia; a Divisão de Publicações registra a publicação de 28 Notas de Física, no período 1956-1957; apresenta também planos para a aquisição de uma máquina de escrever do tipo Varityper, de uma impressora tipo Multilith e o contrato de uma datilógrafa especializada, tudo na direção da modernização dos meios de trabalho, baseados até então na impressão por mimeógrafo, em folhas de Stencil; essa modalidade de reprodução é sumamente trabalhosa, além de pobre do ponto de vista da apresentação, sobretudo para fórmulas e desenhos.

• Além do Monitor a Nêutrons com o qual garantiria a participação do CBPF no Ano Geofísico Internacional, como de fato o fez, o Departamento de Raios Cósmicos apresentava atividades na linha da meteorologia. Uma delas vale a pena ser relembrada, pelo destaque que reassumiu em tempos recentes: trata-se de uma colaboração com a Universidade do Novo México, EUA, sobre a detecção e dosagem do Ozônio na atmosfera. A depleção da camada de Ozônio já fora notada em meados daquela década, mas nenhuma injeção de CFC na atmosfera poderia ser invocada para justificá-la. É bem possível que outros fatores intervenham, além do famigerado gás.

• O Prof. *Ugo Camerini* licenciou-se para uma visita à Universidade de Wisconsin. Os afastamentos de *Lattes* e *Camerini* produziram um vazio muito grande no Departamento de Física Experimental, afetando todas as atividades, principalmente as da Divisão de Raios Cósmicos, inclusive em Chacaltaya.

1957: Ano do Sputnik

• Após 3 de outubro de 1957 as madrugadas não foram mais tão silenciosas: um inusitado bip-bip devassava o silêncio sideral, informando as posições do primeiro satélite artificial da Terra. Foi um feito de grande repercussão. Para os países em desenvolvimento, que buscavam afirmar-se na ciência e na tecnologia, foi uma espécie de referendum das medidas que vinham tomando com grandes sacrifícios mas com renovada esperança. A URSS, que vivera o horror da 2ª Guerra dentro de suas fronteiras, com enormes perdas em patrimônio e vidas humanas, erguia-se diante do mundo, primeiro alcançando os EUA na corrida atômica, depois tomando a frente na ciência espacial, através de um programa bem gerenciado e continuado de educação em todos os níveis e de sólidos investimentos em ciência e tecnologia que pareciam indicar os caminhos da redenção do atraso e da miséria. Dentro dos EUA o Sputnik teve até maior impacto do que *Joe I*;[25] foi criada a NASA para absorver os três projetos espaciais de cada uma das forças armadas americanas que competiam entre si e um esforço muito grande foi feito para garantir a formação de pessoal de alto nível, em números comparáveis aos produzidos nas universidades e institutos soviéticos especializados. Esse propósito foi colocado em prática com tal vontade e eficiência que o número de PhD's formados entre 1960 e 1970 superou de muito a média histórica. No Brasil, que já despertara para a ciência espacial, criando o Centro Tecnológico de Aeronáutica, o Instituto Tecnológico de Aeronáutica, no início da década de '50 e o Instituto Nacional de Pesquisas Espaciais pouco mais tarde, mas ingressava num período de profundas dificuldades políticas e econômicas, a repercussão ficou mais no plano simbólico; ainda assim foi muito estimulante para todos aqueles que empenhavam

---

[25] Como os americanos chamaram a primeira bomba atômica soviética. Alusão a *Joseph Stalin*.

esforços para vencer as diferentes faces do subdesenvolvimento.

• A despeito das dificuldades, o relatório 1957-1958 menciona muitos eventos auspiciosos, sendo talvez o mais importante deles a manutenção do tempo integral e a concessão de um número razoável de bolsas a estudantes brasileiros e bolivianos. O Departamento de Ensino menciona a concessão de bolsas Dulcídio Pereira a dois estudantes brasileiros de Engenharia e duas outras para estudantes bolivianos; também menciona a disponibilidade de bolsas para estudantes do curso de física da FNFi, doze das quais foram concedidas no período. No total, o Departamento de Ensino registra a presença de treze bolsistas e cinco estagiários, durante o período.

• Os trabalhos sobre a radiação cósmica prosseguiam no Rio, com o Monitor a Nêutrons, que já passava à sua fase de produção, e em Chacaltaya, relacionados também ao Ano Geofísico Internacional; entretanto o financiamento das atividades em Chacaltaya vinha ano a ano se tornando mais problemático e o relatório 57/58 anuncia o progresso de um convênio tríplice entre o CBPF, o CNPq e a CNEN para garantir a continuação daquelas atividades.

• Outro grande problema agravado ano a ano foi o da insuficiência de espaço no Mário de Almeida para acomodar todas as atividades. Dois grandes passos foram dados em 1957: a) um convênio com a CNEN para a construção do prédio n° 27, hoje abrigando a *Ciência Hoje*; b) lançamento do edital de concorrência para a construção de nova sede para o CBPF: trata-se do prédio até recentemente ocupado pelo LNCC, que ficou pronto na década seguinte.

• A biblioteca do CBPF relaciona a disponibilidade de 6 462 volumes, entre livros e revistas especializadas.

• A Divisão de Publicações anuncia a ultimação dos preparativos para a publicação da Apostila de Eletrodinâmica Clássica de *J. Leite Lopes*.

• O Departamento do Museu de Ciência continuou à espera de uma definição da Prefeitura do Rio para dar início à sua instalação, totalmente inviável face às carências de espaço na sede do CBPF.

1958: Ano Geofísico Internacional

A comunidade científica internacional adotou esse ano para a realização de trabalhos que pudessem ampliar conhecimentos físicos sobre o planeta, crosta terrestre, atmosfera, magnetismo *etc*. Uma parte importante foi a das variações temporais da radiação cósmica incidente, cujo estudo levou à instalação de uma rede de Monitores a Nêutrons em diferentes latitudes. O CBPF se fez representar com um monitor, construído inteiramente por *Georges Schwaccheim*, com a ajuda de apenas um técnico em eletrônica e um auxiliar de manutenção. A construção começou em fins de '56. O monitor começou a operar dentro das previsões do ano geofísico; funcionou ainda mais um ano e pouco acumulando dados, mas teve de ser desativado logo no início dos anos '60 sob o peso da mais absoluta falta de condições. Mas garantiu a inserção da latitude do Rio de Janeiro no artigo do Handbuch der Physik que tratou do assunto.[26]

• O prof. *Andrea Wataghin* transferiu-se do Departamento de Chacaltaya para a Divisão de Emulsões Nucleares, no Rio. Essa mesma Divisão contou com um professor visitante britânico, *D.J. Prowse*.

• O relatório anuncia ainda a ultimação de medidas para a instalação do acelerador Cockcroft-Walton em dependência do Instituto Militar de Engenharia.

• A despeito das dificuldades, a produção científica dos diferentes setores do CBPF manteve-se elevada; o Departamento de Ensino manteve oito cursos em andamento, além das práticas de laboratório e dezoito estudantes entre estagiários e bolsistas. A biblioteca acusa a compra de 146 livros, a assinatura de 75 títulos de revistas, dez livros doados e 50 novas inscrições de leitores, num período de dois anos.

O relatório assinala com destaque a publicação pelo "Ao Livro Técnico Ltda." do livro Introdução à Teoria Atômica da Matéria e, pelos Cadernos de Cultura do MEC, da monografia "Einstein e Outros Ensaios", ambos de *José Leite Lopes*. Anteriormente, em 1956, aparecia a Física para a Escola Secundária, tradução do livro de *Blackwood-Osgood*, em colaboração com *Jayme Tiomno*. Embora esta iniciativa tivesse tido também importante repercussão no campo do ensino da física no Brasil, a literatura para o ensino em nível médio não era tão carente de títulos em português quanto no caso do nível superior. Aqui o recurso a textos em idioma estrangeiro era inevitável. Às dificuldades idiomáticas e outras, adicionava-se o preço e o desgosto de ser pago em moeda estrangeira; tratava-se aparentemente da *única* mercadoria, entre tantas outras importadas, vinhos, conservas, bacalhau, carros, perfumes, cotada em moeda estrangeira e, pior ainda, para a qual o câmbio não se fazia pelos índices normais, mas bem acima deles. Essa perversão sobre a importação de cultura continua, aparentemente, em curso e quem quer que se coloque no combate ao subdesenvolvimento colide com ela inevitavelmente. O livro do professor *Leite Lopes* foi assim recebido com a dupla satisfação de quem ganha um texto de valor didático e científico e desata um elo apertado na cadeia da subordinação.

---

[26] E. Schopper, E. Lohrmann, G. Mauck, "Neutronen in der Atmosphäre", *Handbuch der Physik* **XLVI/2**, 1967.

1959: Ano do Incêndio

• A 23 de maio de 1959 o CBPF despertou com a notícia de um incêndio que destruíra grande parte de sua biblioteca e a quase totalidade da Divisão de Emulsões Nucleares. A perda total com o sinistro foi avaliada em Cr$50.000.000,00, entre livros, revistas, cerca de dezoito microscópios de pesquisas, placas de emulsões nucleares e outros materiais de trabalho. O problema do espaço no CBPF que já era grande agravou-se, com a destruição das dependências onde se localizavam aquelas unidades. Foram muitos meses dedicados à reconstrução e à seleção de material ainda aproveitável. Quem visitar o setor de periódicos antigos encontrará revistas que não foram consumidas pelo fogo mas onde a água dos bombeiros deixou marcas que o tempo não apagou.

• Também foi um momento de manifestação de grande solidariedade: o CNPq e a CNEN prontificaram-se a conceder recursos para a compra de microscópios e para a reposição de parte do material bibliográfico; a Câmara de Vereadores do Rio e a Câmara Federal de Deputados manifestaram-se também solidárias. Igualmente tocante foi a solidariedade de *Richard Feynman*,[27] que mobilizou companheiros para obter doações individuais de livros e revistas e pelo contato inicial com a Fundação Ford que finalmente doou a maior parte do material bibliográfico que ainda podia ser comercializada.

• Os salários pagos pelo CBPF começavam a mergulhar no nível de alarma; Termos de Convênio começavam a ser assinados com o CNPq para estipendiar pesquisadores no Departamento de Física Teórica e de Radioquímica; o relatório anual, em sua introdução assim se manifesta:

> Como nos anos anteriores, as duas maiores dificuldades encontradas, para a realização do programa de pesquisas foram a falta de pessoal técnico-científico e a deficiência das atuais instalações. Evidentemente estas e outras dificuldades encontradas derivam da exiguidade das verbas disponíveis que continua entravando a ampliação dos trabalhos do Centro. Os salários continuam de baixo nível, desencorajando jovens pesquisadores a prosseguir em suas carreiras científicas, atraídos, como são, por outras instituições que oferecem melhores salários. Por igual motivo torna-se proibitiva a vinda de professores estrangeiros.

• Com toda a dificuldade a produção científica não caiu, nem os cursos oferecidos diminuíram em número, nem o número de estagiários e bolsistas sofreu qualquer diminuição significativa. O problema do espaço encontrou uma solução satisfatória no novo edifício do CBPF, construído em terreno próprio, contíguo àquele em que se localiza o Mário de Almeida; mas isto só aconteceu na década seguinte. O problema de pessoal se agravou na medida em que os salários reais

---

[27] J. Leite Lopes, *Ciência e Sociedade* 013/88.

decresciam continuamente sob o peso da inflação e também na medida que novas alternativas apareciam no país, com a criação do Instituto de Física da PUC, a organização da Universidade de Brasília e, mais para o final dos anos '60, do Instituto de Física da UFRJ e do IFGW, em Campinas. Os salários pagos pelo CBPF só vieram a se recuperar no fim da década de '60, com a criação do FUNTEC, no BNDE.

• De 27 de junho a 7 de julho realizou-se a Escola Latino-Americana de Física, congregando mais de setenta físicos do Brasil, dos países de língua espanhola e de outras partes do mundo. A Escola nasceu de um entendimento entre *Marcos Moshinsky, José Leite Lopes* e do saudoso *J.J. Giambiagi*, durante um encontro científico no México. Desse entendimento nasceu a ideia de institucionalizar o evento, realizando-o rotativamente no México, no Rio e em Buenos Aires. Desde então a Escola se reúne e é um dos eventos científicos mais significativos da comunidade latino-americana. Foi também a semente geradora do CLAF (Centro Latino-Americano de Física).

Comentário Final

• Logo depois que o CBPF completou 21 anos publiquei um ensaio sobre toda a atividade até então realizada, cobrindo portanto os primeiros dez anos.[28] Trata-se de uma análise métrica do comportamento da instituição refletido na evolução temporal dos trabalhos científicos realizados, pessoal, orçamentos, recursos bibliográficos e outros de infraestrutura, buscando também algumas correlações entre característicos internos e destes com parâmetros exteriores à instituição. Àquele texto, no interesse da contenção deste relato já demasiado extenso, remeto o leitor mais curioso. Aqui importa apenas mencionar que a grande contribuição do CBPF em seus primeiros dez anos foi seu pioneirismo na reforma universitária. Integrando a pesquisa científica ao ensino superior o CBPF antecipou-se em cerca de 20 anos ao restante do país, que só o veio fazer dentro dos quadros da pós-graduação, novamente com o CBPF na liderança. Em que pese o valor e atualidade dos trabalhos científicos aqui realizados, nada ultrapassou, em termos de combate ao subdesenvolvimento, as lições sobre tempo integral, modernização das práticas do ensino e formação de pessoal qualificado que, através das crises sucessivas, saiu do CBPF para ocupar posições na universidade brasileira, levando a experiência adquirida. Do quadro de coordenadores da Universidade Nacional de Brasília fizeram parte dois professores do CBPF e o primeiro Diretor do Instituto Central de Ciências foi também egresso de seus quadros. A formação de pessoal para o projeto nuclear em seus primeiros anos não foi menos importante; nos órgãos técnicos da CNEN, no Instituto de Radiação e Dosimetria, no IEN encontramos

---

[28] A. Marques, "CBPF: 21 Anos de Atividades Científicas", *Ciência e Sociedade*, 1971.

expressivo contingente de físicos formados pelo CBPF em seus primeiros anos, tendo também membros saídos de seus quadros de pesquisa ocupado por duas vezes a Presidência da CNEN.

• Espera-se aqui, após longo e heterogêneo relato, uma palavra de síntese que configure a linha mediana da trajetória do CBPF. Ao menos a questão da resistência à extinção, após tantas crises e carências continuadas, se demanda esclarecer. Confesso, após compilar os dados dos documentos e organizá-los neste relato, mesmo com a vantagem de tê-los testemunhado quase todos, que me senti incapaz de qualquer conclusão significativa, engessado na perplexidade resultante da simples recordação deles à luz de parâmetros que pudessem explicar aquela resistência. Entrego-os, assim, à reflexão dos leitores, na esperança de que sua isenção os faça mais livres para refletir, avaliar e concluir. Deixo apenas o registro de que, na busca de respostas simples, atribuí aquela resistência primeiro a um grupo limitado de pessoas, que progressivamente fui estendendo até terminar com a longa lista de pessoal arrolado no anexo 11. Pessoas comuns que eram, em sua esmagadora maioria, foram, entretanto, incomuns na dedicação, tendo dado à instituição muito mais do que tiraram dela. Esta talvez seja a grande lição a extrair dos dez primeiros anos do CBPF. É possível que a conjuntura, dominada por fortes vetores ideológicos, tenha colaborado, dando a cada um a sensação de que seu trabalho ultrapassava o significado no plano pessoal para projetar-se nos grandes meta-projetos em curso na época: a reforma universitária e o projeto nuclear. Quem sabe?

ANEXOS

1949: Ano da Fundação

Assembléia Geral:

A Assembleia Geral do CBPF foi constituída pelos Membros Fundadores, aqueles que assinaram a Ata de fundação, estendido esse título a outras personalidades que, mesmo não tendo assinado aquela Ata tiveram essa formalidade dispensada pelos Membros Fundadores e Membros Efetivos, aqueles que foram substituindo os fundadores pelo falecimento ou qualquer outro impedimento. O Relatório 1949-1950 alinha os seguintes Membros Fundadores:

Abrahão de Moraes, Adalberto Menezes de Oliveira, Agostinho Jaensch, Aluísio Bezerra Coutinho, Álvaro Alberto da Mota e Silva, Amaury Menezes, Anibal Fernandes, Anísio Teixeira, Antonio Aniceto Monteiro, Antonio Bezerra Baltar, Antonio Carlos Teixeira, Antonio José da Costa Nunes, Augusto Araújo Lopes Zamith,

Benedito Castrucci, Candido Lima da Silva Dias, Elza Gomide, Edison Farah, Fernando Furquim de Almeida, Humberto Grande, João da Silva Monteiro, José Otavio Knaak de Souza, Luiz Henrique, Jacy Monteiro, Leopoldo Nachbin, Omar Catunda, Antonio Rodrigues, Armando Dubois Ferreira, Arthur Moses, Arthur Hehl Neiva, Ary Nunes Tiethbolt, Augusto Frederico Schmidt, Bernardino C. de Mattos Netto, Bernardo Gross, Blandina Azeredo Fialho, Branca Fialho, Caio Libanio de Noronha Soares, Carlos Chagas Filho, Cesar Guinle, Cesar Lattes, Cristóvão Colombo dos Santos, Cyrillo Hercules Florence, Dulcídio A. Pereira, Eduardo de Macedo Soares e Silva, Eduardo Schmidt Monteiro de Castro, Elisa Frota Pessoa, Elisiario Tavora Filho, Elsa Cesario Alvim, Ernani da Mota Rezendo, Ernesto Luiz de Oliveira Junior, Euvaldo Lodi, Francisco Clementino Santiago Dantas, Francisco Magalhães Gomes, Francisco Matarazzo Sobrinho, Francisco Mendes de Oliveira Castro, Francisco Xavier Roser S.J., Gabriel Emiliano de Almeida Fialho, Geraldo Rocha Lima, Henry British Lins de Barros, Hervasio Guimarães de Carvalho, Homero Barbosa de Assis Martins, Hugo Ribeiro, Jayme Tiomno, Jean Meyer, João Alberto Lins de Barros, João Consoni Perrone, João Cristovam Cardoso, João Holmes Sobrinho, Joaquim da Costa Ribeiro, Joaquim de Faria Góes Filho, Jorge Americano, Jorge de Oliveira Castroi, José Carneiro Felipe, José Leite Lopes, José Moreira dos Santos Pena, Josué Lage, Lauro Xavier Nepomuceno, Lelio Itapoambira Gama, Lino Leal de Sá Pereira, Lourenço Borges, Luiz de Barros Freire, Luiz Cintra do Prado, Luiz Osório de Siqueira Netto, Luiz Paes Leme, Luiz Soroa Filho, Maria Laura Moura Mousinho, Mario Alves Guimarães, Mario Camarinha da Silva, Mario Henrique Bettanio de Azevedo, Mario Werneck, Martha Siqueira Lattes, Mauricio Mattos Peixoto, Miguel Mauricio da Rocha, Moacyr Teixeira da Silva, Nelson Alberto Lins de Barros, Nelson Chaves, Newton da Silva Maia, Orlando Rangel Sobrinho, Oromar Moreira, Oswaldo Aranha, Oswaldo Frota Pessoa, Oswaldo Gonçalves de Lima, Othon Henri Leonardos, Paulino Cavalcanti, Paulo de Assis Ribeiro, Paulo Berredo Carneiro, Paulo Emidio de Freitas Barbosa, Paulo Ribeiro de Arruda, Paulo Saraiva Toledo, Paulus Aulus Pompéia, Petronio de Almeida Magalhães, Roberto Marinho de Azevedo, Roberto Maurell Lobo Pereira, Roberto Aureliano Salmeron, Romildo Pessoa, Rômulo Barreto de Almeida, Teófilo Alvares da Silva, Walter de Camargo Schutzer.

Diretoria:

Presidente: *Ministro João Alberto Lins de Barros;* Vice-Presidente: *Alte. Álvaro Alberto da Mota e Silva;* Diretor Científico: *Professor Cesar Lattes;* Diretor Tesoureiro: *Cmte. Gabriel Fialho;* Diretor Executivo: *Dr. Paulo de Assis Ribeiro.*

Direção Técnica:

Prof. Carlos Chagas Filho; Prof. Francisco Mendes de Oliveira Castro; Prof. Joaquim da Costa Ribeiro; Prof. José Leite Lopes; Prof. Luiz Cintra do Prado.

Conselho:

Cel. Armando Dubois; Prof. Artur Moses; Cel. Bernardino C. de Matos Neto; Prof. Ernani da Mota Rezende; Prof. José Carneiro Felipe; Prof. Lelio Itapoambira Gama; Prof. Lino Leal de Sá Pereira; Cel. Orlando Rangel Sobrinho; Prof. Paulo Ribeiro de Arruda.

Lista de Adesões:

Não consta qualquer registro nominal de funcionários no período. Os cientistas, técnicos
e funcionários administrativos, quando remunerados pelo CBPF, receberam por serviços prestados, sem qualquer contrato. Consta, entretanto, do Relatório Anual 1949-1950 uma lista de personalidades que contribuíram com quantias diversas, somando um milhão duzentos e cinquenta e nove mil quinhentos e um cruzeiros e cinquenta centavos, para a manutenção da instituição. É a seguinte a relação de nomes e entidades:

Mário de Almeida, Serviço Social da Indústria, SESI, Anônimas, Guilherme Guinle, Paulo de Assis Ribeiro, Escritório Técnico Paulo de Assis Ribeiro, Metalúrgica Matarazzo SA, Usinas Pernambucanas, Renato Soeiro, Armando Queiroz Monteiro, Departamento de Física da Universidade Católica, Manoel M. Batista da Silva, Peixoto de Castro, Rivadávia Correia Meyer, Usina Sto. Inácio SA, Usina Catende, Mirsilo Gasparri, Romeu V. Queiroz, Expansão Mercantil, Horácio Saldanha, Pedro da Cunha, Edmundo Barreto Pinto, Presidente Eurico Gaspar Dutra, Gen. Góes Monteiro, Governador Moisés Lupion, Benedito Valadares, Juscelino Kubitschek, Pereira Lira, Eduardo Gomes, Horácio Lafer, Alcides Moutinho Neiva, Alice Flexa Ribeiro, Amintas Jacques de Moraes, Antonio Bezerra Cavalcanti, Argemiro Couto de Barros, Branca Fialho, Beraldo Melo, Carvalho & Cia., Cia. Mineira de Várias Indústrias, Djalma C. Fontes, Eduardo Maia Franco, Elza Schneider, Francisco Mendes de Oliveira Castro, Jaime Queiroz Monteiro, João Colares Moreira, Joaquim de Oliveira Sampaio, José Brito Passos, José do Nascimento Brito, José Pinto Carvalho Osório, Josué de Castro, Mário Botti, Léo Amaral Pena, Mário Moutinho Neiva, Nanto Ribeiro Junqueira, Nicolino Malleta, Nino Galo, Newton Freitas de Souza, Newton Silva Maia, Renato Onofre P. Aleixo, União Católica, Vera Assis Ribeiro, Walter Lemos Azevedo, Wilfrido Shorto.

Ano de 1950

Diretoria:

Presidente: *Ministro João Alberto Lins de Barros;* Vice-Presidente: *Alte. Álvaro Alberto da Mota e Silva*; Diretor Científico: *Professor Cesar Lattes;* Diretor Tesoureiro: *Cmte. Gabriel Fialho;* Diretor Executivo: *Nelson Lins de Barros.*

Direção Técnica:

Prof. Carlos Chagas Filho; Prof. Francisco Mendes de Oliveira Castro; Prof. Joaquim da Costa Ribeiro; Prof. José Leite Lopes; Prof. Luiz Cintra do Prado.

Ano de 1951:

Não dispomos de documentos. Tampouco dispomos de qualquer indicação sobre mudanças nos órgãos de direção.

Ano de 1952

Diretoria:

Presidente: *Ministro João Alberto Lins de Barros;* Vice-Presidente: *Alte. Álvaro Alberto da Mota e Silva;* Diretor Científico: *Professor Cesar Lattes;* Diretor Tesoureiro: *Cmte. Gabriel Fialho;* Diretor Executivo: *Professor Hervasio de Carvalho.*

Direção Técnica:

Prof. Carlos Chagas Filho; Prof. Antonio José da Costa Nunes; Prof. Joaquim da Costa Ribeiro; Prof. José Leite Lopes; Prof. Luiz Cintra do Prado.

Conselho:

Cel. Armando Dubois; Prof. Artur Moses; Cel. Bernardino C. de Matos Neto; Prof. Ernani da Mota Rezende; Dr. João Carlos Vital; Prof. Lelio Itapoambira Gama; Prof. Lino Leal de Sá Pereira; Cel. Orlando Rangel Sobrinho; Prof. Paulo Ribeiro de Arruda.

Ano de 1953

Diretoria:

Presidente: *Ministro João Alberto Lins de Barros*; Vice-Presidente: *Alte. Álvaro Alberto da Mota e Silva*; Diretor Científico: *Professor Cesar Lattes;* Diretor Executivo: *Álvaro Difini.*

Conselho Deliberativo:

Arthur Moses (presidente), Roberto Marinho de Azevedo, Augusto Araújo Lopes Zamith, Jorge de Oliveira Castro, Armando Dubois Ferreira, Carlos Chagas Filho, Elisiário Távora, Lélio I. Gama, Humberto Grande, Bernardino de Mattos Neto, Antonio José da Costa Nunes, Adalberto Menezes de Oliveira, Dulcidio Pereira, Roberto Maurell Lobo Pereira, Ernani da Mota Rezende, Joaquim da Costa Ribeiro, Orlando F. Rangel Sobrinho, Anysio Teixeira.

Conselho Técnico-Científico:

*Cesar M.G. Lattes (presidente), *Guido Beck, Ugo Camerini, F.M. de Oliveira Castro, Ismael Escobar, J. Leite Lopes, Leopoldo Nachbin, Jayme Tiomno.*

Secretário Geral:

*Nelson Lins de Barros.

Período 1954-1955

Aqui os relatórios anuais cobrem períodos de junho a julho do ano seguinte.

Diretoria:

Presidente: *Gal Edmundo de Macedo Soares e Silva*, Vice-Presidente: *Cap. de Frag. Henry British Lins de Barros*, Diretor Científico: *Prof. Francisco Mendes de Oliveira Castro*, Diretor Executivo: *José Machado de Faria*, Secretário Geral: *Nelson Lins de Barros*.

Conselho Deliberativo:

Prof. Antonio José da Costa Nunes (Presidente), Cel. Armando Dubois Ferreira, Profa. Maria Laura Mousinho, Prof. Anysio Spínola Teixeira, Prof. Carlos Chagas Filho, Prof. Lelio Itapoambira Gama, Cap. de Corv. Roberto Maurell Lobo Pereira, Prof. Ernani da Mota Rezende, Dr. Jorge Mendes de Oliveira Castro.

Conselho Técnico-Científico:

Prof. Francisco Mendes de Oliveira Castro (Presidente), Prof. Cesar Lattes, Prof. Guido Beck, Prof. Hervasio Guimarães de Carvalho, Prof. Ismael Escobar, Prof. Jayme Tiomno, Prof. José Leite Lopes, Prof. Leopoldo Nachbin, Prof. Ugo Camerini.

Período 1955-1956

Diretoria:

Presidente: *Gen. Edmundo de Macedo Soares e Silva*, Vice-Presidente: *Cap. de Frag. Henry British Lins de Barros*, Diretor Científico: *Prof. Francisco Mendes de Oliveira Castro*, Diretor Executivo: *José Machado de Faria*, Secretário Geral: *Nelson Lins de Barros*.

Conselho Deliberativo:

Prof. Antonio José da Costa Nunes (Presidente), Cel. Armando Dubois Ferreira, Profa. Maria Laura Mousinho, Prof. Anysio Spínola Teixeira, Dr. João Consoni Perrone, Prof. Lelio Itapoambira Gama, Cmte. Gabriel Emiliano de Almeida Fialho, Prof. Ernani da Mota Rezende, Dr. Jorge Mendes de Oliveira Castro.

Conselho Técnico-Científico:

Prof. Francisco Mendes de Oliveira Castro (Presidente), Prof. Cesar Lattes, Prof. Guido Beck, Prof. Hervasio Guimarães de Carvalho, Prof. Ismael Escobar, Prof. Jayme Tiomno, Prof. José Leite Lopes, Prof. Leopoldo Nachbin, Prof. Luís Marquez, Prof. Ugo Camerini.

Período 1956-1957

Diretoria:

Presidente: Gen. Edmundo de Macedo Soares e Silva, Vice-Presidente: Prof. Antonio José da Costa Nunes, Diretor Científico: Prof. Francisco Mendes de Oliveira Castro, Diretor Executivo: José Machado de Faria, Secretário Geral: Nelson Lins de Barros.

Conselho Deliberativo:

Prof. Ernani da Mota Rezende (Presidente), Alte. Adalberto Menezes de Oliveira, Prof. Anysio Spínola Teixeira, Dr. Antonio Carlos Barbosa Teixeira, Prof. Augusto Araújo Lopes Zamith, Sra. Blandina de Azeredo Fialho, Cmte. Gabriel Emiliano de Almeida Fialho, Dr. João Consoni Perrone, Dr. Jorge Mendes de Oliveira Castro.

Conselho Técnico-Científico:

Prof. Francisco Mendes de Oliveira Castro (Presidente), Prof. Cesar Lattes, Prof. Guido Beck, Prof. Hervasio Guimarães de Carvalho, Prof. Ismael Escobar, Prof. Jayme Tiomno, Prof. José Leite Lopes, Prof. Leopoldo Nachbin, Prof. Luís Marquez, Prof. Ugo Camerini.

Período 1957-1958

Diretoria:

Presidente: Gen. Edmundo de Macedo Soares e Silva, Vice-Presidente: Prof. Antonio José da Costa Nunes, Diretor Científico: Prof. Francisco Mendes de Oliveira

Castro, Diretor Executivo: *José Machado de Faria*, Secretário Geral: *Nelson Lins de Barros*.

Conselho Deliberativo:

Prof. Ernani da Mota Rezende (presidente), Alte. Adalberto Menezes de Oliveira, Prof. Anysio Spínola Teixeira, Dr. Antonio Carlos Barbosa Teixeira, Prof. Augusto Araújo Lopes Zamith, Sra. Blandina Azeredo Fialho, Cmte. Gabriel Emiliano de Almeida Fialho, Dr. João Consoni Perrone, Dr. Jorge Mendes de Oliveira Castro.

Conselho Técnico-Científico:

Prof. Francisco Mendes de Oliveira Castro (Presidente), Prof. Cesar Lattes, Prof. Guido Beck, Prof. Hervasio Guimarães de Carvalho, Prof. Ismael Escobar, Prof. Jayme Tiomno, Prof. José Leite Lopes, Prof. Leopoldo Nachbin, Prof. Luís Marquez, Prof. Ugo Camerini.

Período 1958-1959

Diretoria:

Presidente: *Gen. Edmundo de Macedo Soares e Silva*, Vice-Presidente: *Prof. Antonio José da Costa Nunes*, Diretor Científico: *Prof. Guido Beck*, Diretor Executivo: *José Machado de Faria*, Secretário Geral: *Nelson Lins de Barros*.

Conselho Deliberativo:

Prof. Ernani da Mota Rezende (presidente), Alte. Adalberto Menezes de Oliveira, Prof. Anysio Spínola Teixeira, Dr. Antonio Carlos Barbosa Teixeira, Prof. Augusto Araújo Lopes Zamith, Sra. Blandina Azeredo Fialho, Cmte. Gabriel Emiliano de Almeida Fialho, Dr. Jorge Mendes de Oliveira Castro, Prof. Paulo Emidio Barbosa.

Conselho Técnico-Científico:

Prof. Guido Beck (Presidente), Prof. Francisco Mendes de Oliveira Castro, Prof. Cesar Lattes, Prof. Hervasio Guimarães de Carvalho, Prof. Ismael Escobar, Prof.

Jayme Tiomno, Prof. José Leite Lopes, Prof. Leopoldo Nachbin, Prof. Luís Marquez, Prof. Ugo Camerini.

Pessoal

Estão listadas as pessoas que, a qualquer título, prestaram serviços continuados ao CBPF e cujos nomes foram registrados nos relatórios e outros documentos oficiais do período 1949-1959. Exceções foram feitas àqueles que exerceram apenas mandatos em colegiados diversos e aos Professores Visitantes por prazos inferiores a um ano, já mencionados no texto. Às pessoas do sexo feminino que mudaram o nome em consequência de matrimônio foi adicionado entre colchetes o nome de casada.

Adel da Silveira, Agostinho David, Agostinho Lage Ornellas de Souza, Alberto B. Madureira, Alberto de Carvalho Peixoto de Azevedo, Alceu Gonçalves do Pinho Filho, Aguiar Guimarães, Alcidia de Jesus, Aldízio F. Costa, Alfredo Hendel, Alfredo Marques, Alfredo da Silva Bento, Alfredo Ventura da Costa, Alice Gonçalves Barbosa [Rivera], Aloysio Nunes Costa, Amaury Alves Menezes, Andrea Wataghin, Anna Christina Ferreira Teixeira [Monteiro de Barros], Anna Maria Martins de Oliveira Freire [Endler], Annita Mischan [Macedo], Antonio Aniceto Monteiro, Anselmo Salles Paschoa, Antonio Luciano Leite Videira, Antonio Francisco V. Seixas, Antonio P. Cabral, Antonio A. Miranda, Antonio José Duffles de Andrade Amarante, Antonio Maria Meira Chaves, Argus F.O. Moreira, Arthur Gerbasi da Silva, Arthur Luiz de Amorim Nóbrega, Assir Américo dos Santos, Atanael Ferreira Lopes, Benedito da Costa Carvalho, Benicio Grossi, Blandina Fialho, Carlos Alberto Dias, Carlos Alberto Heras, Carlos Lopes Conceição, Carlos Alberto Magno da Silva, Carlos Marcio do Amaral, Carmen Marques Pereira Wohrle, Cecilia M. Kotin, Cesar Lattes, Chaim Samuel Hönig, Claudio Arcoverde Leal de Barros, Claudio Basbaum, Claudio P. Monteiro de Barros, Clotilde Zuleta Bilbao, Colber Gonçalves de Oliveira, Conceição do Nascimento, Danilo Marcondes, David Gorodovits, Delia Valerio Ferreira, Dilson Ribeiro de Almeida, Dulce Fabiano Rodrigues Gomes, Dulce Rielo [Moura Filho], Edisonina Vieira Vaz, Edith A. Saldanha, Edmar Ferdinando Sixel, Eduardo Styzei, Eduardo Maldonado, Elde Pires Braga, Elena Dionê Borgli Brandão, Elisa Frota Pessoa, Elisa Van Tol, Elisabeth Luis Pinto, Erasmo Madureira Ferreira, Erich Willner, Ernesto Leitão, Ernesto Passarelli, Eugenio Lerner, Eugenio Lopes Faria, Eugênio Trombini Pellerano, Ewa Wanda Cybulska, Fanor Rocha, Fernando Gomide, Fernando de Souza Barros, Flavio dos Santos Carvalho, Francis D. Murnaghan, Francisco de Arruda Camargo, Francisco Brandão, Francisco Calheiros, Francisco Mendes de Oliveira Castro, Francisco Theodoro, Francisco Viana Freitas, Gabriel Fialho, Frima Kastansky, George Rawitscher, Georges Schwachheim, Geraldo Alves dos Santos, Geraldo Americo dos Santos, Geraldo Arnoldi Pedrozo,

213

Gerhard Hepp, Gert Molière, Giuseppe Occhialini, Guido Beck, Guilherme Fontes Leal Ferreira, Gustavo A. Garnier, Hans Joos, Harold Cordeiro Oest, Helcio da Silva Prallon, Helio Nazario Severo Leal, Helmut Schwarz, Henry British Lins de Barros, Herman Glanz, Herch Moyses Nussenzweig, Hervasio Guimarães de Carvalho, Hildebrandes Antonio da Silva, Hilton Vieira, Homero Lima Brandão, Homero Alcides Brandão Viegas, Horacio Cintra de Magalhães Macedo, Hugo Chaves Moreira, Ismael Escobar Vallejo, Ivan Otero Ribeiro, Jacques A. Danon, Jayme Maifeld, Jayme Tiomno, Joaquim Duarte Pinto, Joaquim Santiago, Joaquim Jerônymo de Moura Filho, João Carlos Fernandes, João Tavares da Silva, Jesualdo José Valente, Jorge Luís Barros Reis, Jorge Alberto Barroso, Jorge Americo Süssman, Jorge André Swiecka, Jorge Wilson Alves, José Alves Feitosa, José Augusto Viana, José Leite Lopes, José A. Lutterbach, José Arthur Borges Cabral, José Fabio Neto, J.J. Giambiagi, José de Lima Acioly, José Luís Correia Vieira, José Nogueira, José Pereira de Andrade, José Rocha, Josias Leandro de Oliveira, Josué Ferreira da Silva, Joviano C. Valadares, Julio Fernandes Gomes Filho, Jorge Mendes Santana, Juvenal Xavier de Oliveira, Juan Hersil, Klaus Stefan Tausk, Laércio Gondim de Freitas, Laudelino Azeredo de Souza, Lauro Xavier Nepomuceno, Leda Araujo de Moura [Srivastava], Léa Távora de Miranda Bastos, Lelé Ribeiro Gil, Léo Câmara Neiva, Leon Jordan, Leonora Carneiro Felipe, Leopoldo Nachbin, Leroy Schwarz, Lidia dos Santos, Lindolpho de Carvalho Dias, Luís Alves Menezes, Luís Marquez y Bajos, Luís Fonseca Tavares, Luiz José de Moura, Luiz Machado de Lima, Luiz Felipe Villares Paiva, Luiz Carlos Gomes, Maria dos Anjos Martins Cid Lopes, Magin Zubieta, Manfred Beurlen, Manoel Cavalheiro, Manoel Porto, Mauricio Mattos Peixoto, Maria Sylvia L. Alves, Marilda Gosling Velloso, Maria Cecilia dos Santos, Maria Decimar Martins, Maria da Gloria Rios, Maria Laura Mousinho [Leite Lopes] Maria do Rosário Magri, Maria Yvone da Rocha Barroso, Mario Amoroso Anastácio, Mario Henrique Simonsen, Mario Silvestre, Mario Marchesini dos Santos, Marlene Bonacossa Mello, Martha Maria McDowell Brito Pereira, Maria Luiza Cordeiro Gerk, Marisa Ballariny, Matilde Sette Ferreira Pires, Mauricio José das Chagas Machado, Michael Malagolowkin, Micheline Claire Levi [Nussenzweig], Miguel Curi, Milton José, Myrian Pereira da Silva, Moacyr da Silva, Nair Miranda, Nara Pereira Terra, Nelson Lins de Barros, Neusa Margem [Amato], Neusa da Silva Pinto, Neusa de Souza Alho, Newton Braga, Newton Castanheira Brandão, Neyla Leal da Costa, Nelson Leite Durães, Nicim Zagury, Nilton Nascimento, Nilza Eny de Freitas Almeida, Nilza de Jesus, Newton Castanheira, Oscar Troncoso Lozada, Olga Margem, Omar Serrano de Abreu, Osorio Chagas Meirelles, Oswaldo Costa, Oswaldo Fernandes de Oliveira, Otília Pinheiro Ribeiro de Castro, Paulo Emidio Barbosa, Paulo Jorge da Silva, Paulo Poppe de Figueiredo, Paulo Ribemboim, Pedro Luís van Tol Filho, Peter K. Weyl, Prem Prakash Srivastava, Raimundo A. Normando, Raymundo Correa Gomes, Raymundo Gonçalves de Paiva, Regina de Araujo Góes [Jakubowicz], Reginaldo dos Santos, Regis Alves da Silva Aisikovitch, Ricardo de Carvalho Ferreira, Ricardo Escobar Vallejo, Ricardo Palmeira, Richard Feynman, Richard Miller, Roberto Bastos da Costa, Roberto Claudio das Neves Leitão, Roberto Fernandes de Mello, Roberto

*Moreira Xavier de Araújo, Roberto Nicolsky, Roberto Pimentel, Roberto Aureliano
Salmeron, Rubens Paiva, Rudolph Charles Thom, Ruth Rodrigues, Ruth Senra,
Samuell Wallace McDowell, Sarah de Castro Barbosa [Andrade], Sergej Lebedev,
Sergio Telles Ribeiro, Siegfried Oschalins, Silvio da Silva Moura, Solange May
Cuyabano [de Barros], Suely Gaertner, Suzana L. de Souza Barros, Sylvio Gomes,
Teófilo Marinho da Cruz, Teresinha Medeiros, Teresinha T. Villar, Ugo Camerini,
Vania Junqueira Monteiro de Barros, Violeta de Jesus Couto Gomes, Waldir Perez,
Walter F. Moura, Willy de Souza, Wilson Luiz de Góes, Wilson Manoel dos Santos,
Yara dos Santos Chaudon, Yvonne Barra [Lerner], Ximenes Alexandrino da Silva,
Zaída Meirelles, Zenir Calil da Silva.*

REUNIÃO FINAL — ENCERRAMENTO DO "SYMPOSIUM" SÔBRE RAIOS CÓSMICOS — Rio de Janeiro, 8 de agôsto de 1941

1 — G. Wataghin
2 — Donald Hughes
3 — Norman Hilberry
4 — Arthur Moses
5 — Arthur H. Compton
6 — William P. Jesse
7 — Ernest O. Wollan
8 — René Wurmser
9 — Francisco Souza
10 — F. M. de Oliveira Castro
11 — F. Venancio Filho
12 — I. Costa Ribeiro
13 — Othon Nogueira
14 — F. Magalhães Gomes
15 — Arthur do Prado
16 — Alvaro Alberto
17 — Menezes de Oliveira
18 — Junqueira Schmidt
19 — Yolande Monteux
20 — Paulo R. Arruda
21 — G. Occhialini
22 — M. Cruz
23 — Carlos Chagas Jr.
24 — Ignacio Azevedo Amaral
25 — M. D. de Souza Santos
26 — B. Gross
27 — Abrahão de Morais
28 — Paulus A. Pompeia
29 — Pe. F. X. Roser S.J.

Figura 17.1: Participantes do Simpósio sobre Radiação Cósmica, Rio de Janeiro, agosto de 1941, patrocinado pela Academia Brasileira de Ciências. Primeira reunião internacional da Física realizada no Brasil. A numeração de identificação dos participantes flui da esquerda para a direita na primeira fileira e no sentido inverso pra os que estão de pé, a partir do número 10.

Über die Entstehung von Radiumisotopen aus Uran durch Bestrahlen mit schnellen und verlangsamten Neutronen[1].

Von O. Hahn und F. Strassmann, Berlin-Dahlem[2].

In einer Reihe von Arbeiten haben Meitner, Hahn und Strassmann[3] die Vorgänge aufgeklärt, die bei der Bestrahlung des Urans mit Neutronen zu Elementen mit höherer Ordnungszahl als 92, also zu den sog. Transuranen führen. Außer drei künstlichen Uranisotopen werden 6 Transurane nachgewiesen und in ihren chemischen Eigenschaften festgestellt. Sie gehören zu den Elementen 93—96. In jüngster Zeit wurde von denselben Verfassern nach ein 60-Tage-Körper aufgefunden[4], der vermutlich ebenfalls ein Transuran ist, dessen Stellung aber in den drei isomeren Reihen bisher noch nicht ganz sichergestellt ist.

In mehreren Mitteilungen befassen sich I. Curie und P. Savitch[5] mit einer weiteren Substanz, die sie bei der Bestrahlung des Urans mit Neutronen erhalten haben; sie schreiben ihr eine Halbwertszeit von $3^1/_2$ Stunden zu, mit chemischen Eigenschaften, die bisher nicht genau festzustellen waren. Der Körper wurde von Curie und Savitch zuerst für ein Thorisotop gehalten, später für ein Actiniumisotop, dann aber festgestellt, daß er weder das eine noch das andere ist, sondern vermutlich ein Transuran, aber mit Eigenschaften, die von den von Hahn, Meitner und Strassmann festgestellten Eigenschaften der Transurane in bemerkenswerter Weise verschieden sein. Curie und Savitch diskutieren mehrere Möglichkeiten, die ihnen aber selbst schwer verständlich und unbefriedigend vorkommen:

1. Der 3,5-Stunden-Körper hat die Ordnungszahl 93, und die bisher nachgewiesenen Transurane haben statt der Kernladungen 93—96 die Kernladungen 94—97.

2. Der 3,5-Stunden-Körper hat die Ordnungszahl 94, und die bisher beschriebenen Transurane haben die Ordnungszahlen 93, 95—97.

3. Der 3,5-Stunden-Körper ist isomer zu einem der bekannten Transurane, hat aber eine abweichende Elektronenanordnung, so daß trotz gleicher Kernladung mit einem normalen Transuran die chemischen Eigenschaften einer seltenen Erde auftreten.

Zu 1. ist zu sagen, daß diese Annahme nicht zutreffen kann, weil über die chemische Natur der von Meitner, Hahn und Strassmann beschriebenen Eka-Rheniumisotope, also des Elements 93, kein Zweifel bestehen kann.

Zu 2. Abgesehen von den abweichenden chemischen Eigenschaften hätte sich ein 3,5-Stunden-Körper als Zwischenprodukt zwischen den Elementen 93 und 95 an den Aktivitätskurven der Transurane bemerkbar machen müssen.

Zu 3. Diese Annahme ist äußerst unwahrscheinlich, wie ja Curie und Savitch auch betonen. Die auch in absolut unwägbarer Menge vorliegenden anderen künstlichen Radioelemente haben sich chemisch immer so verhalten, wie man aus ihrer Stellung im periodischen System erwarten sollte.

Bei Gelegenheit neuer Versuche über die chemischen Eigenschaften der Transurane haben wir deshalb versucht, auch den Curie-Savitchschen 3,5-Stunden-Körper nachzuweisen. Es ist uns auch gelungen, die Substanz nach der von den Verff. angegebenen Abscheidungs- und Meßmethode zu erhalten.

Die genauere Prüfung führte zu bemerkenswerten Ergebnissen. Diese sollen an dieser Stelle nur kurz zusammenfassend dargestellt werden und sind in den Zahlenangaben noch als vorläufig anzusehen.

Bei der Bestrahlung des Urans mit Neutronen entstehen vermutlich drei isomere Radiumisotope, die also durch zwei sukzessive α-Umwandlungen über Thorium entstanden sein müssen. Daß es sich dabei um Radiumisotope handelt und daß sie nicht etwa aus dem unbestrahlten Uran stammen, wurde nach mehreren Methoden einwandfrei bewiesen. Ihre Halbwertszeiten sind ungefähr 25 Min., 110 Min., mehrere Tage.

Aus diesen isomeren Radiumisotopen entstehen durch β-Strahlenemission drei isomere Actiniumisotope, die als solche nachgewiesen wurden. Ihre Halbwertszeiten sind, vorerst in roher Annäherung, ungefähr mit 40 Min., 4 und 60 Stunden anzugeben. Aus diesen Actiniumisotopen entstehen vermutlich drei Thorisotope, über die aber bisher noch nichts auszusagen ist.

Da die Substanzen wohl alle das Atomgewicht 231 haben, und da es unter den natürlichen Radioelementen bereits ein Thorisotop vom Atomgewicht 231, nämlich UY mit etwa 28 Stunden Halbwertszeit gibt, wird es von Interesse sein, nachzuweisen, ob eines der drei genannten künstlichen Thorisotope mit UY identisch ist.

Was den von Curie und Savitch beschriebenen 3,5-Stunden-Körper betrifft, so glauben wir, daß er ein Gemisch der von uns im einzelnen nachgewiesenen und chemisch identifizierten Körper vorstellt. Die von den Verfassern angegebenen Eigenschaften ihres 3,5-Stunden-Körpers sind mit den Eigenschaften eines solchen Gemisches durchaus verträglich. In ihrer letzten Mitteilung weisen übrigens die Verfasser darauf hin, daß ihr 3,5-Stunden-Körper augenscheinlich noch weitere Substanzen längerer Halbwertszeit enthält, von denen aber bisher weder Halbwertszeiten noch genetische Zugehörigkeit feststellbar seien.

Obgleich es sich bei dem zu den Radiumisotopen führenden Prozeß primär um eine α-Strahlenabspaltung

---

[1] Eingegangen am 8. November 1938.

[2] Kaiser Wilhelm-Institut für Chemie.

[3] Z. B. O. Hahn, L. Meitner u. F. Strassmann, Ber. dtsch. chem. Ges. 70, 1374 (1937). — L. Meitner, O. Hahn u. F. Strassmann, Z. Physik 106, 249 (1937).

[4] O. Hahn, L. Meitner u. F. Strassmann, Naturwiss. 26, 475 (1938).

[5] I. Curie u. P. Savitch, J. Physique et Radium 8, 385 (1937) — C. r. Acad. Sci. Paris 206, 906, 1643 (1938) — J. Physique et Radium 9, 355 (1938).

Figura 17.2: Página 755 do *Naturwissenschaften* 46 de dezembro de 1938, onde *Hahn* e *Strassmann* anunciaram a formação de elementos de número atômico inferiores ao do Urânio, Rádio e possivelmente Tório, além de transurânicos, após a irradiação de Urânio com nêutrons. Logo a seguir os autores complementaram esse resultado mostrando que parte dos radioisótopos identificados como Rádio eram isótopos de Bário. *Lise Meitner* e *Otto Frisch* logo interpretaram o resultado deste trabalho como a fissão dos núcleos de Urânio. Este trabalho não foi consequência nem foi determinante de qualquer mudança de paradigma na Física, mas depois dele mudou o modo de produção de toda a Ciência e, de certo modo, mudou o Mundo.

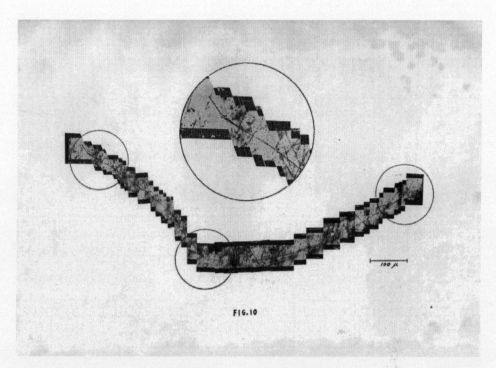

Figura 17.3: Do primeiro trabalho experimental realizado integralmente no CBPF: um estudo sobre o modo eletromagnético de desintegração do méson π^+, publicado em *Anais da Academia Brasileira de Ciências* XXII, nº 4 (1950) por *Elisa Frota Pessoa* e *Neusa Margem*. As emulsões foram irradiadas em Berkeley por *Lattes* e examinadas em microscópicos emprestados do Instituto de Química Agrícola e do Departamento de Polícia Técnica.

Figura 17.4: Chegada de *Leite Lopes* ao Rio (1950). A foto mostra parte expressiva da liderança científica e administrativa do CBPF. Na fila da frente, da esquerda para a direita: *Martha* e *Cesar Lattes, Leite Lopes, Nelson Lins de Barros, Carmita*, esposa de *Leite, Blandina Fialho, D. Marieta,* sogra de *Leite,* com o neto *Sérgio* no colo. Na fileira de trás, da esquerda para a direita: *Gabriel Fialho, Adel da Silveira, Oliveira Castro* e dois não identificados.

Figura 17.5: Aspectos do Laboratório de Alto-Vácuo (acima) e da Biblioteca (abaixo). Reproduzidas de "Notícia", CBPF, 1953.

Figura 17.6: Aspectos dos Laboratórios de Desenvolvimento (acima) e de Montagens Eletrônicas (abaixo). Reproduzidas de "Notícia", CBPF, 1953.

Figura 17.7: *Eduardo Stizey* no Torno para Vidros. Reproduzida do Relatório Anual 1955-56.

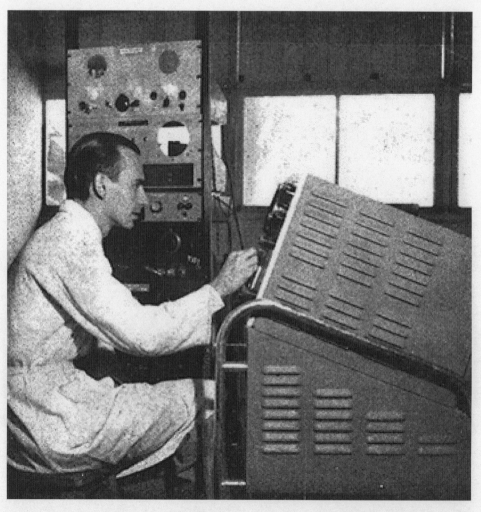

Figura 17.8: *Georges Schwachheim* no processo de montagem do Monitor a Nêutrons.

Figura 17.9: Pavilhão Mário de Almeida. Foto: Luis Lima, CBPF, 1955.

APÊNDICE I[29]

P.S. Se quiserem cartas mais legíveis e mais cuidadosamente redigidas arranje-se o dinheiro para uma secretária.

Berkeley, Janeiro 10, 1949

Caro Leite,

Acabo de lhe enviar longo telegrama. Aqui vão os detalhes:

Quando contei ao Lawrence s novidades daí, nossas possibilidades e principalmente quando soube da ajuda do João Alberto e de quem o João Alberto é e pode fazer, o "cidadão" ficou entusiasmadíssimo. Reuniu os "Big Shots" para resolver sobre a melhor maneira de ajudar-nos.

Ele acha melhor, ao invés de comprarmos uma alta tensão, trazermos para cá imediatamente três engenheiros eletrotécnicos ou dois e um físico (o Jean Mayer seria ideal) para que aprendam logo a fazer um ciclotron. Acha ele que em três meses podemos ter um pequeno ciclotron de 2 MeV (prótons) ou 1 MeV (dêuterons) funcionando. Aqui construíram um de 2 MeV por 15 000 dólares em 6 semanas; já está funcionando perfeitamente e está sendo usado como modelo do de 10 MeV que será construído para injeção no bevatron. Os engenheiros tomariam parte na construção do de 10 MeV.

Quanto à construção do nosso, temos duas possibilidades:

1) Entrar em entendimento com o Fabio Collins para que o ciclotron seja construído por eles. Os nossos engenheiros iriam participar da construção.

2) Caso o Lawrence consiga a permissão da comissão E.A. [de Energia Atômica] para no ajudar oficialmente é quase certo que poderemos construí-lo aqui. Então será uma mamata.

O Lawrence foi a Washington e estará de volta dentro de uma semana. Ele ia falar com o Lilienthal sobre ajuda para nós e disse-me que tem certeza de que a conseguirá pois o tipo ficou muito satisfeito com a história do méson artificial. Eles estão interessados em mostrar que a C.E.A. está patrocinando pesquisa pura e séria, uma boa oportunidade para mostrar aos demais países que eles estão dispostos a ajudar *etc. etc.*

O Lawrence acha que seria muito útil o João Alberto vir logo aos E.U. Gostaria de tê-lo aqui para uma curta visita para conhecer a organização do laboratório e discutir os problemas e dificuldades que encontraremos devido às nossas

---

[29] Cópia de carta de *Cesar Lattes a J. Leite Lopes* detalhando entendimento havidos com *Ernest O. Lawrence* para a participação de um grupo de brasileiros na construção de um pequeno ciclotron que poderia ser trazido para o Brasil a preço de custo.

condições locais. A ideia inicial do Lawrence era de convidar o João Alberto "to keep him interested" pois ele tinha medo que o João Alberto não tivesse compreendido bem nossos problemas e poderia ser útil ele vir para cá e ficar de certo modo comprometido com o pessoal daqui. Quando expliquei melhor a atitude do João ele compreendeu que isso não seria necessário mas, assim mesmo, acha que seria útil o João vir e entrar em contato com o modo de trabalho de um laboratório grande como este e possivelmente ficar conhecendo outros laborat[órios] (eu poderia acompanhá-lo) + gente como o Oppy e o Raby que poderão nos ajudar em outras coisas. O Lawrence pede, pois, que eu transmita o convite informar (*sic*) ao Ministro e Senhora. Diz o Lawrence que se encarregará também de "entertain" ambos. Não creio que pense em pagar a[s] estadias isso seria pedir muito ⋯ Mostre a carta ao João (ao qual peço desculpas por ter tirado o título para economia de espaço e tempo) e respondam logo.

Quanto aos engenheiros nossos, sejam bem cuidadosos na escolha. O Lawrence acha que quanto mais moços melhor. Eu concordo mas acho que um deveria ter uma certa experiência não só técnica como administrativa; seria muito bom se o Paulo Arruda pudesse vir. Com o João, caso haja dinheiro, talvez conviesse vir o Cintra (ele fala inglês?) ou o Assiz Ribeiro (caso ele seja o nosso diretor administrativo).

Dr. Cookey, o diretor administrativo daqui, recomenda que ao pedirem o visto, os eventuais "ciclotron-makers" mencionem "*para trabalhar no ciclotron de Berkeley*". Isso significa que o cônsul pedirá conselho ao Departamento de estado; este à Comissão da Energia Atômica e esta já estará informada pelo Lawrence de tudo de modo que "every thing will do smoothly". É importante que isso seja feito assim pois por praxe o Departamento de Estado deve ser o primeiro a ser informado. Talvez, a juízo do João e do consul Cross, seria bom este último escrever uma carta ao Lawrence dizendo qualquer coisa como: "Que está informado de nossos planos e conhece o pessoal e acha (se acha) que é gente de confiança e amigos dos Estados Unidos ou coisa que valha."

Uma possibilidade que é ainda bem remota mas que existe é a seguinte: Já está bem adiantada a construção de um modelo de escala 1/4 do bevatron de 6 MeV aqui. Eles vão usá-lo para estudar no modelo todos os problemas que surgirão no grande de modo que encontrarão solução sem gastar muito dinheiro. Pois bem: Daqui a uns dois anos o modelo será completamente inútil para a turma daqui e provavelmente seria desmantelado. Foi-me sugerido por vários colegas daqui (não big shots infelizmente) que procurasse entrar em entendimento com o Lawrence para ver se não seria possível comprar o bevatronzinho quando este se tornar inútil ao laboratório. Provavelmente custará algo como 200.000,00 (duzentos mil dólares) o que é muito dinheiro mas, quem sabe se em dois anos não poderemos nos permitir o luxo disso? O bevatronzinho pode dar (uma vez adaptado) meio milhão (quinhentos milhões) de volts (prótons); a intensidade será pequena mas adiante da energia"who cares"?

Não tive ainda tempo de trabalhar muito nos mésons. Temos tido dificuldades em obter boas exposições pois o feixe não está ainda bem regulado e vive mudando de plano. Nas melhores exposições obtivemos dez vezes mais mésons para o mesmo background do que sem as alpha[s]. Isto para mésons de pequena energia. Deve haver um número muito grande de mésons de alta energia mas ainda não podemos estudá-los. Wilson Powell, um com estrela, já os teve seis na câmara de Wilson; o trabalho dele está parado porque a câmara explodiu-lhe na cara. Felizmente não sofreu ferimentos sérios. Gardner continua no hospital; teve um pneumotórax natural e quase morre. Passou duas semanas em tenda de oxigênio mas está melhor. Só amanhã conseguirei ir vê-lo pois tive que obter autorização do médico para visitá-lo. Acho que ele está fora de uso pelo menos por mais dois meses.

Nada mais no memento. Toquem para a frente a sociedade e comecem a arranjar a gaita pois precisamos aproveitar enquanto está quente. Um abraço

(assinado) Cesar

Caso queiram telefonar meu telefone é olimpic 32830 (*casa* estou sempre a noite depois de 12 p.m. podem chamar a qualquer hora:
Está um frio incrível.
Vários graus abaixo de zero! California!

APÊNDICE II[30]

ACORDOS FIRMADOS ENTRE C.B.P.F. E OUTRAS ENTIDADES

> *Termo pelo qual o Centro Brasileiro de Pesquisas Físicas, sociedade civil, com sede nesta capital, à Rua Álvaro Alvim, 21, 21° andar, e a Universidade do Brasil, órgão autônomo, na conformidade do Decreto-lei n° 8.393, de 17 de dezembro de 1945, convencionam o seguinte:*

1°) O Centro Brasileiro de Pesquisas Físicas fará construir, por sua própria conta, um pavilhão, em terreno cedido a título precário, pela Universidade do Brasil, com prévia aprovação do Ministro da Educação e Saúde, para ali instalar a sua sede, pavilhão este com a especificação e de acordo com o desenho constante da planta anexa.

---

[30] Acordo com a Universidade do Brasil para a cessão do terreno onde foi edificado o Pavilhão Mário de Almeida.

2°) O referido pavilhão destina-se à sede do Centro Brasileiro de Pesquisas Físicas, que instalará nele os seus laboratórios, salas de aula, biblioteca, e demais dependências necessárias aos estudos e pesquisas físicas e matemáticas, abrangendo a cooperação com o ensino destas especialidades ministrado pela Universidade do Brasil.

3°) No caso de dissolução do Centro Brasileiro de Pesquisas Físicas, ou da Fundação com idênticos fins, prevista no artigo 18, § 2°, dos Estatutos respectivos, passará à propriedade da União, como parte integrante do próprio nacional ora ocupado pela Universidade do Brasil, o pavilhão a que se refere a cláusula primeira, para ser utilizado em benefício do ensino científico, ou da melhor forma, a juízo do Governo da União.

Nada a mais havendo a acrescentar e por estarem de pelo acordo com todas as condições estabelecidas no presente termo, subscrevem o mesmo como legítimos representantes das entidades mencionadas:

Doutor *Pedro Calmon Muniz de Bittencourt* pela Universidade do Brasil.

Professor *Cesar Lattes*, pelo Centro Brasileiro de Pesquisas Físicas.

Rio de Janeiro, 18 de maio de 1950.

Como testemunhas: *Antônio Carneiro Leão. – Joaquim da Costa Ribeiro.*

APÊNDICE III[31]

ACORDOS FIRMADOS ENTRE CENTRO BRASILEIRO DE PESQUISAS FÍSICAS E OUTRAS ENTIDADES

> *Termo do acordo entre a Universidade do Brasil e o Centro Brasileiro de Pesquisas Física[s], para outorga de mandato universitário ao segundo, nos termos do art. 3° do Decreto-lei n° 8.393, de 17 de dezembro de 1945, combinado com o § 1° do art. 8° do Estatuto da Universidade.*

A Universidade do Brasil, entidade autônoma por força de decreto-lei n° 8.393, de 17 de dezembro de 1945, representada pelo Reitor, professor Deolindo Augusto de Nunes Couto, usando da faculdade que lhe confere o art. 8°, § 1° do Estatuto respectivo, aprovado pelo Decreto n° 21.321 de 18 de junho de 1946, e na conformidade da autorização do Conselho Universitário, consoante do processo n° 11.014/49, e o Centro Brasileiro de Pesquisas Físicas, sociedade civil com sede

---

[31] Acordo firmado entre a Universidade do Brasil e o CBPF pelo qual lhe foi outorgado Mandato Universitário.

nesta Capital, representada pelo seu Diretor Científico, Prof. Cesar Lattes, resolvem firmar o presente acordo, mediante o qual é outorgado, ao segundo, mandato universitário, sob as condições seguintes:

1ª O Centro Brasileiro de Pesquisas Físicas obriga-se a colaborar com a Universidade do Brasil, criando ou promovendo cursos especializados, cujos programas serão aprovados pelo Conselho Universitário.

2ª Os Laboratórios do Centro Brasileiro de Pesquisas Físicas poderão ser utilizados pelos professores, técnicos especializados, pesquisadores e estudantes da Universidade do Brasil, obedecendo a programas de trabalho previamente aprovados pelas duas partes contratantes.

3ª O pessoal científico e técnico do Centro Brasileiro de Pesquisas Físicas prestará à Universidade do Brasil a necessária colaboração nos objetivos previstos nas cláusulas 1ª e 2ª.

4ª A Universidade do Brasil obriga-se a reconhecer os cursos promovidos pelo Centro Brasileiro de Pesquisas Físicas, na conformidade da cláusula 1ª, e a expedir certificado de aprovação aos alunos que os tenham frequentado, com aproveitamento devidamente apurado em provas e exames.

E, por estarem de pleno acordo, firmam o presente termo as duas partes, na presença das testemunhas abaixo.

Rio de Janeiro, 11 de agosto de 1950. – *Deolindo Couto.* – *Cesar Lattes.*

Testemunhas: *João Alberto Lins de Barros – Álvaro Alberto da Mota e Silva – Joaquim da Costa Ribeiro – José Leite Lopes – Gabriel Fialho – Nelson Luis de Barros – Ernani da Mota Resende – Francisco Mendes de Oliveira Castro.*

APÊNDICE IV[32]

Acordo de colaboração entre a Escola Nacional de Química e o Centro Brasileiro de Pesquisas Físicas.

A Escola Nacional de Química e o Centro Brasileiro de Pesquisas Físicas, representado respectivamente pelos Diretor e Presidente, como suplementação ao acordo estabelecido a 11 de agosto de 1950 entre a Universidade do Brasil e o Centro Brasileiro de Pesquisas Físicas, em virtude do qual foi outorgado mandato universitário ao último, e com base no item *h* do artigo 3° dos Estatutos do referido

---

[32] Em complementação ao Mandato Universitário, foram firmados convênios com a Escola Nacional de Química (aqui reproduzido), com a Escola Nacional de Engenharia e com a Faculdade Nacional de Filosofia tornando regimentais os cursos e créditos concedidos.

Centro, firmam o presente, respeitadas as cláusulas do mencionado mandato.

1°) O Centro Brasileiro de Pesquisas Físicas e a Escola Nacional de Química, colaborarão, na medida de suas possibilidades, sem ônus para quaisquer das partes, na realização de cursos e programas de pesquisas de interesse comum.

2°) Os laboratórios e bibliotecas de cada entidade poderão ser frequentados por professores, alunos, bolsistas e estagiários, devidamente credenciados, da outra entidade, sem prejuízo de seus serviços normais.

3°) O Centro Brasileiro de Pesquisas Físicas porá à disposição da Escola Nacional de Química os serviços de deu Departamento Técnico, de acordo com as normas do mesmo, e cobrado os preços de custo.

Rio de Janeiro, 29 de dezembro de 1956. – Prof. *Annibal Cardoso Bittencourt.* Diretor da Escola Nacional de Química.

18

Cosmic Ray Physics and the Very Large Hadron Collider

Alberto Santoro

High Energy Physics has always been in the frontier of Science and Technology. The discoveries of quarks, like the Top Quark [1] and others, have been part of a systematic effort to understand the Standard Model, the fundamental interactions and their constituents. Many laboratories were built to make these goals possible. We have observed that the higher the energy and luminosity, the easier to find new particles. Building new machines demands the development of new technologies in several domains.

Recently the SSC (Super Superconducting Collider) was terminated. The ideas behind this project remain alive though, and we still have some initiatives of the same magnitude, like the Eloisatron project, coordinated by Prof. Zichichi in Italy [2].

In the present, we have a periodic conference dedicated to the ideas around a Very Large Hadron Collider (VLHC) to which I had the honor of being invited to participate and give a talk [3]. This conference is aimed at discussing the progress in physics and the necessary technologies to be employed in a possible future machine. One of the ideas is to build a 200 TeV machine with perimeter of approximately 300 km.

The discussions about the internationalism of Science have also been the subject of reflection [4]. Certainly, the 100 to 1 000 TeV region falls within the domain of Cosmic Ray Physics (C.R.P.). We will try to show how we can relate topics of C.R.P. to diffractive production processes. This is the range that can be named the Real Beyond Standard Model (RBSM) or the New Standard Model (NSM). Certainly, this will be the physical region of the new physics where the all-asymptotic theorems could be tested [$\sigma(pp) = [\sigma(p\bar{p})$, e.g.]. But why talk about these subjects when we would like to talk about Cesare Mansueto Giulio Lattes? This is indeed an easy question with a clear answer. Lattes was always in the vanguard of Particle Physics and his ideas will be not be exhausted in this century, as we will demonstrate below. We have many examples of particle discoveries in Cosmic Ray; among them we can quote the following discoveries:

- positron and muon [5]

- π-meson [6]

- Charmed Particles [7]

- Except for the *tau* lepton, and bottom and top quarks, all other quarks and leptons were first observed in Cosmic Ray Physics.

All these discoveries have been confirmed by many different experiments in what one can call Accelerator Physics. However, I would like to emphasize some of the most exciting events in Particle Physics: the Centauro events, discovered by Lattes and his collaborators [first reported by Lattes in the 13th ICRC (Denver) **4**, 267 (1973)] [8].

A Centauro event is defined as a Cosmic Ray event with a low electromagnetic activity, or the absence of gammas originated from π^0 and a large number of hadron decays. Its main characteristics are:

- It was observed in Nuclear Collisions at 100-1 000 TeV energy range.

- The production has about 100 hadrons.

- No gamma from π^0-decays appear.

- All hadrons with an average transverse momentum p_t=1–2 GeV.

- The production was about 50 m from the interaction point.

Cosmic Ray Physics have contributed to the development of Particles Physics not only with new particle discoveries but also with techniques of detection like streamer chambers, multicoincidence systems, and so on. In the present there are many collaborations taking into consideration Cosmic Rays results and working together.

Let us assume now that the objects like Centauros make part of the physics in the new range scale 100–1 000 TeV.

In this scale of energy the Diffractive Physics phenomena will be very important. The corresponding diffractive mass that can be probe in both present and future collider machines is estimated by the formula

$$M_x^2(1-x)$$

Where s is the total center of mass energy squared and x is the fraction of the momentum carried by the diffracted proton or anti-proton (see the following table for the diffractive mass in some colliders).

Tevatron -	LHC -	VLHC
$\sqrt{s} = 2.$ TeV	$\sqrt{s} = 14.$ TeV	$\sqrt{s} = 100.$ TeV
	Single Diffraction	
$M_x = 450.$ GeV	$M_x = 3.6$ TeV	$M_x = 22.5$ TeV
	Double Diffraction	
$M_x = 100.$ GeV	$M_x = 800$ GeV	$M_x = 5.$ TeV

Now we see how important these values are for Diffractive Physics. As the Centauro masses are around a few hundreds of GeV and Mini-Centauro family about tens of GeV, in principle, we have the capability to observe it at DØ detectors in the Tevatron energies. Particularly in DØ we have recently proposed a set of new detectors that will increase the possibility of observing diffractive events. The insertion of these detectors will make more feasible the observation of Centauro events in accelerators.

Our hopes are also based in the values for the diffractive mass available at Tevatron. And it is interesting to see that the values of diffractive mass for the different processes (single and double Pomeron exchange) increase with the energy. Then, it is possible to see that soon we will have detectors to tag the proton and the anti-proton diffracted in the p p̄ collisions at the Tevatron. We will guarantee the characteristic of diffraction of the event and trigger the produced jets in a convenient way. We have, as we saw above, enough energy to produce Centauros.

The Large Hadron Collider (LHC) is expected to start operating around 2005, opening a new era for Diffractive Physics and a new perspective for the observation of Centauros. As diffraction does not disappear at very big energy, and perhaps is even the dominant mechanism to produce these events, it will be very interesting to continue working in the Centauro's Physics. One can guess that at the LHC and VLHC energies Centauros will be produced at least by diffractive processes. If this hypothesis becomes true, one can expect the birth of "new physics".

We have suggested that Monte Carlo simulations for diffractive production of Centauros should be initiated immediately.

Finally, I would like to stress, again, the importance and impact of Lattes' ideas in present day particle physics. More than allow the observation of Centauros in accelerators, I believe that the next generation of machines will shed light on some of the most intriguing mysteries of the microcosmo.

I would like to acknowledge that these notes were written motivated by the kind invitation of my colleagues F. Caruso, A. Marques, and A. Troper, which I accepted with a great pleasure. I felt honored by the opportunity to write about this great Brazilian physicist, Cesare Mansueto Giulio Lattes.

References

[1] S. Abachi *et al.* – DØ Collaboration, *Phys. Rev. Lett.* **74**, 2422 (1995).

[2] Proceedings of the "Hadron Colliders at the Highest Energy and Luminosity" – Ed. by A.G. Ruggiero – World Scientific (1996) and references therein.

[3] Collections of notes and talks of the Very Large Hadron Collider Physics and Detector Whorshop. March 13-15 (1997).

[4] The practice of the internationalism in High Energy Physics is an old idea. Everything is showed by several cultures separating several countries around one well defined purpose: "to probe the matter in all its aspects".

[5] C.D. Anderson, *Phys. Rev.* **43**, 491 (1933); S.H. Neddermeyer and C.D. Anderson, *Phys. Rev.* **51**, 884 (1937).

[6] C.M.G. Lattes, G.P.S. Occhialini and C.F. Powell, *Nature* **160**, 453 (1947) and Part II, *ibid.* p. 486; C.M.G. Lattes, H. Muirhead, G.P.S. Occhialini and C.F. Powell, *Nature* **159**, 694 (1947).

[7] K. Nin *et al. Prog. Theo. Phys.* **47**, 280 (1972).

[8] S. Abachi *et al.* – DØ Collaboration – *Phys. Rev. Lett.* **74**, 2422 (1995).

19

O Méson pi em Bristol Hoje

Cássio Leite Vieira

Num dos andares do Royal Fort, edifício que abriga o Laboratório H.H. Wills, da Universidade de Bristol, passa quase despercebida a presença de uma caixa de madeira clara sobre uma mesa, cercada por três painéis com fotos e textos. A caixa protege um velho (porém, bem-conservado) Cooke, Troughton & Simms M408, microscópio que resiste com bravura às, no mínimo, cinco décadas que recaem sobre ele.

A princípio, nada parece chamar a atenção para aquele canto do corredor, e as poucas pessoas que passam pelo local o fazem sem maior interesse pelas peças ali depositadas. No entanto, o veterano Cooke é o último sobrevivente de uma das batalhas mais emocionantes que a física deste século travou: a detecção do méson pi, partícula responsável por carregar a chamada quarta força da natureza, a força forte, que mantém os prótons coesos no núcleo atômico.

Como admirador desse período da história da física, gosto de pensar que pelas duas objetivas daquele velho microscópio passou o olhar atento do físico inglês Cecil Frank Powell (1903-1969), prêmio Nobel de 1950. Ou talvez o do físico italiano Giuseppe Occhialini (1907-1993), ou dos então jovens físicos brasileiros Cesar Lattes e Ugo Camerini.

A missão de quem olhava pelas lentes desses microscópios era encontrar, em meio a uma selva de milhares de diminutos traços emaranhados, a trajetória de uma nova partícula. Era um trabalho cansativo que demandava horas e horas para esmiuçar uma única chapa de emulsão nuclear. Mas a recompensa, como sabemos hoje, foi gratificante para a equipe de Bristol: no início de 1947, a microscopista

Figura 19.1: O laboratório H.H. Wills, onde a existência do méson pi foi confirmada em 1947, funciona no quarto andar do prédio Royal Fort, em Bristol (Inglaterra). O edifício é conhecido como "Torre do Cigarro", pois sua construção foi financiada por Henry Herbert Wills, proprietário de uma indústria de tabaco que doou, no início da década de 1920, £200 mil pounds para a construção do H.H. Wills, inaugurado em outubro de 1927. Foto: Alícia Ivanissevich, Bristol (1997).

Figura 19.2: Estação de trem onde Cesar Lattes chegou a Bristol no início de 1946, vindo do porto de Liverpool, no qual desembarcou do Santo Rosário, o primeiro a transportar passageiros do Brasil para a Europa depois do fim da Segunda Guerra Mundial. Foto: Alícia Ivanissevich, Bristol (1997).

Figura 19.3: Vista atual do Laboratório de física H.H. Wills da Universidade de Bristol. O prédio Royal Fort fica atrás desse prédio. Foto: Cássio Leite Vieira, Bristol (1997).

Figura 19.4: Último exemplar no Laboratório H.H. Wills do microscópio Cooke, Troughton & Simms, Ltd (model M 408, 1 div - 1 micron), usado para "escanear" as placas de emulsões nucleares em busca dos traços das partículas, trabalho feito na época por uma equipe de mulheres microscopistas sob a liderança do físico inglês Cecil Frank Powell (1903-1969). Foto: Cássio Leite Vieira, Bristol (1997).

Marietta Kurz observou a primeira imagem do decaimento de um méson pi em méson mi.

Há ainda na caixa de madeira uma outra relíquia dos tempos em que se perscrutava o núcleo atômico com chapas de emulsão, microscópicos e uma paciência de causar inveja a um monge tibetano. Devidamente protegida, uma pequena placa quadrada de vidro, com não mais do que cinco centímetros de lado, é o testemunho aparentemente modesto de uma época em que se fazia física experimental de fronteira com relativamente pouco dinheiro. Foi nesse pedaço de vidro (e, talvez, com a ajuda desse último Cooke) que a recém-formada física Rosemary Brown, uma das microscopistas do H.H. Wills e esposa do físico Peter Fowler, visualizou em 1949 o primeiro decaimento $\pi - \mu - e$. Por sinal, Fowler, neto do grande físico neozelandês Ernest Rutherford (1871-1937), publicou seus primeiros trabalhos, ainda como um graduando em física, juntamente com Lattes (infelizmente, Fowler havia morrido poucos meses antes de minha ida a Bristol, como me contou Rosemary em entrevista).

O astrofísico Rodney Hillier, pesquisador aposentado e ainda professor do Instituto de Física da Universidade de Bristol, age como curador do nosso modesto memorial ao méson pi (entenda-se microscópio e painéis). Com gestos que lembram um cerimonial ensaiado ao longo dos anos, Hillier primeiramente liga o microscópico e faz brilhar novamente a luz azulada do velho Cooke. Pinga uma gotícula de um tipo de óleo sobre a pequena placa de vidro antes de colocá-la sob as lentes do microscópio. Cinquenta anos depois, tem-se a chance de repetir o que os pesquisadores e as microscopistas do H.H. Wills faziam à exaustão (devo confessar que me causou uma certa emoção observar aqueles três "risquinhos" que revelaram ao mundo o primeiro decaimento $\pi-\mu-e$ da história). Na avenida Tyndall, em frente à nova fachada do H.H. Wills, está a biblioteca da Universidade de Bristol. Em seu subsolo, encontra-se o *Wills Memorial Library*, onde estão depositados os arquivos de Powell. Uma busca rápida no arquivo revela duas referências ao nome de Cesar Lattes. Uma delas é uma foto (não inédita) na qual o físico brasileiro posa ao lado da equipe de pesquisadores do H.H. Wills. Nada que pudesse ser considerado um "furo" jornalístico. Mas a segunda referência é preciosa: o diário de laboratório de Lattes, de julho de 1947, em cujas páginas estão, entre outras anotações, os cálculos da massa do méson pi, feitos com base nas chapas que Lattes trouxe do monte Chacaltaya, na Bolívia, ainda em 1947.

O que mais surpreende é que o *Physics Note Book*, com a assinatura "C.M.G. Lattes" na capa, parece ser o único documento relevante do ano em que o méson pi foi detectado – há outros diários, alguns de Powell e Occhialini, mas de anos anteriores ou posteriores ao descobrimento. É um documento importante para entender o método usado por Lattes para calcular a massa da partícula proposta corajosamente pelo físico japonês Hideki Yukawa (1907-1981) em meados da década de 1930. Deixo aqui, então, minha sugestão para algum físico ou historiador da ciência que se proponha a destrinchar as fórmulas, as tabelas e os gráficos do caderno. Hoje, o espaço onde trabalhavam as microscopistas do H.H. Wills foi dividido em gabinetes que estão ocupados por pesquisadores.

A sala escura, na qual as chapas eram reveladas, é atualmente o local onde os cientistas e os funcionários do H.H. Wills enchem seus *mugs* com café, na tentativa de amenizar os efeitos do clima frio e teimosamente chuvoso de Bristol. No chão, ao lado da porta de entrada da ex-sala escura, está um baú velho. Dentro dele, estão caixas de chapas de emulsão nuclear da Ilford e Kodak, usadas na década de 1940. Numa delas, do tipo C2, encontra-se a assinatura "Peter Fowler".

Num salão, no térreo do *Royal Fort*, está um tipo de Galeria dos Notáveis. Alinhados por toda a extensão das paredes, estão retratos emoldurados de pesquisadores famosos que passaram pela Universidade de Bristol. Lá está Lattes, com seus vinte e poucos anos, ao lado de tantos outros físicos notáveis. Na foto, a camisa branca está impecável, e os cabelos pretos penteados à *Hollywood*. A primeira imagem que vem à mente é a de um artista de cinema da década de 1940 e não a de um cientista. Como era de se esperar, o tempo tratou de remodelar o H.H. Wills, varrendo assim quase toda a história do méson pi no laboratório. No entanto, a cerca de um quarteirão do *Royal Fort*, ainda permanece incólume o pub Robin Hood, no qual Lattes, Occhialini, Camerini e outros integrantes do H.H. Wills

Figura 19.5: Dr. Rodney Hillier, astrofísico e professor do Instituto de Física da Universidade de Bristol, instalado andares abaixo do Laboratório H.H. Wills, no Royal Fort. Dr. Hillier é o mantenedor do pouco que resta da história da descoberta do méson pi na "Torre do Cigarro": o microscópio Cooke, Troughton & Simms, Ltd e algumas placas de emulsões nucleares, entre elas a que mostra o primeiro decaimento do méson pi em méson mi e elétron (decaimento $\pi - \mu - e$) de 1949. Foto: Cássio Leite Vieira, Bristol (1997).

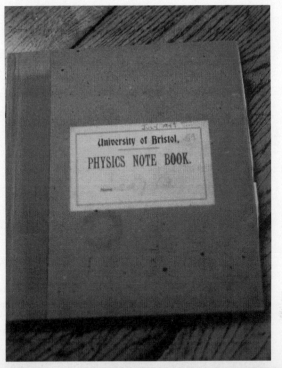
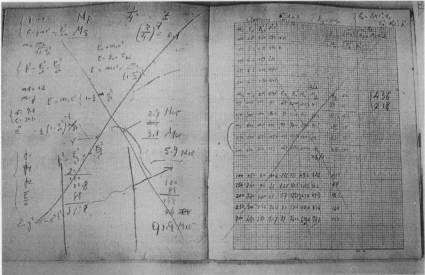

Figura 19.6: Caderno de laboratório (*note book*) usado por Cesar Lattes em julho de 1947. Nele, estão os cálculos que levaram à massa do méson pi depois das exposições feitas no monte Chacaltaya, na Bolívia, poucos meses antes. Depositado na Wills Memorial Library, em Bristol, é o único documento de Lattes nessa biblioteca de documentos raros. Foto: Cássio Leite Vieira, Bristol (1997).

Figura 19.7: Rosemary Fowler (*née* Brown), em sua casa em Bristol, ao lado do mosaico de imagens que indica o primeiro decaimento do méson pi em méson mi e elétron. Física pela Universidade de Bristol, ela foi a primeira microscopista graduada na equipe de Powell. Foi também contemporânea de Lattes durante a estada do brasileiro no laboratório. EM 1949, ela detectou o primeiro decaimento $\pi-\mu-e$ em placas que já eram sensíveis ao elétron. Seu marido, o físico Peter Fowler, falecido recentemente (N.E.: em 1996), publicou com Lattes seus primeiros trabalhos em *Nature* 159, 301 (1947) e *Proc. Roy. Soc.* LIX, 883 (1947). Foto: Cássio Leite Vieira, Bristol (1997).

Figura 19.8: Caixas das placas de emulsões nucleares, fabricadas pelas empresas Ilford e Kodak, usadas no estudo de raios cósmicos pela equipe do Laboratório H.H. Wills por volta do ano do descobrimento do méson pi. Foto: Cássio Leite Vieira, Bristol (1997).

Figura 19.9: Notícia publicada no jornal *Bristol Evening World*, de 9 de abril de 1947, na qual sw lê, no primeiro parágrafo, que o "Dr. C.M.G. Lattes, pesquisador assistente em física atômica na Universidade de Bristol, embarcou em Bristol na última segunda-feira com destino à América do Sul. Sua chegada ao Rio de Janeiro estava prevista para ontem. Ele deverá voar sobre os Andes para expor chapas fotográficas a uma altitude entre 6 mil e 6,3 mil metros.".É interessante notar que Lattes foi agraciado na reportagem com o título de *Doutor Honoria Causa*. O texto afirma ainda que "Lattes deverá voar sobre os Andes", quando, em realidade, as exposições das placas sempre foram planejadas, desde o início, para serem feitas em terra. Foto: Cássio Leite Vieira, Bristol (1997).

Figura 19.10: Cesar Lattes na "Galeria dos Notáveis" no Instituto de Física da Universidade de Bristol. Na legenda da foto, lê-se: "C.M.G. Lattes – 1946-47", anos em que trabalhou no Laboratório H.H. Wills. Foto: Cássio Leite Vieira, Bristol (1997).

buscavam refrescar a garganta com um *pint* da melhor cerveja do mundo antes de voltarem para mais uma jornada (em geral, noturna) de trabalho no laboratório.

Figura 19.11: "The Robin Hood", nas proximidades do prédio Royal Fort, os pesquisadores do H.H. Wills – incluindo Cesar Lattes – aproveitavam para tomas a última cerveja antes de retornarem ao trabalho no laboratório. Na época, o proprietário recebia com prazer os cientistas, mesmo que eles aparecessem poucos minutos antes do horário de fechamento, ainda hoje às 23 h. Foto: Cássio Leite Vieira, Bristol (1997).

Antes de partir de Bristol – portando, lógico, uma cópia do *Note Book* e dois filmes de fotografias –, senti-me na obrigação de cumprir uma última tarefa: entrei no Robin Hood e pedi um *pint* de Guinness (imaginei que essa poderia ser a marca preferida por um dos personagens desta história). Antes do primeiro gole, levantei a caneca e saudei a memória daqueles que, já tendo partido, participaram da descoberta do méson pi, cujos desdobramentos foram tão importantes para a ciência no Brasil e no mundo. E, claro, desejei vinda longa aos que ainda estão entre nós.